Interfacial Transition Zone
in Concrete

JOIN US ON THE INTERNET VIA WWW, GOPHER, FTP OR EMAIL:

WWW: http://www.thomson.com
GOPHER: gopher.thomson.com
FTP: ftp.thomson.com
EMAIL: findit@kiosk.thomson.com

A service of I(T)P

RILEM REPORTS

RILEM Reports are state-of-the-art reports prepared by international technical committees set up by RILEM, The International Union of Testing and Research Laboratories for Materials and Structures. More information about RILEM is given at the back of the book.

RILEM REPORT 11

Interfacial Transition Zone in Concrete

**State-of-the-Art Report prepared by RILEM
Technical Committee 108-ICC,
Interfaces in Cementitious Composites**

RILEM
(The International Union of Testing and Research
Laboratories for Materials and Structures)

Edited by

J. C. Maso
LMDC, Toulouse, France

CRC Press
Taylor & Francis Group
Boca Raton London New York

CRC Press is an imprint of the
Taylor & Francis Group, an **informa** business
A TAYLOR & FRANCIS BOOK

CRC Press
Taylor & Francis Group
6000 Broken Sound Parkway NW, Suite 300
Boca Raton, FL 33487-2742

First issued in paperback 2019

© 1996 RILEM
CRC Press is an imprint of Taylor & Francis Group, an Informa business

No claim to original U.S. Government works

ISBN-13: 978-0-419-20010-9 (hbk)
ISBN-13: 978-0-367-86389-0 (pbk)

A catalogue record for this book is available from the British Library

Publisher's Note

This book has been produced from camera ready copy provided by the individual contributors in order to make the book available for the Workshop.

**Visit the Taylor & Francis Web site at
http://www.taylorandfrancis.com**

**and the CRC Press Web site at
http://www.crcpress.com**

Contents

Contributors – RILEM Technical Committee 108-ICC

M.G. Alexander Department of Civil Engineering, University of Cape Town,
Rondebosch, South Africa

A. Bentur National Building Research Institute, Faculty of Civil Engineering,
Technion, Israel Institute of Technology, Haifa, Israel

J.C. Maso Laboratoire Matériaux et Durabilité des Constructions,
(Chairman) I.N.S.A.-U.P.S. Génie Civil, Complexe Scientifique de Rangueil,
31077 Toulouse, France

M. Massat Laboratoire Matériaux et Durabilité des Constructions, Complexe
Scientifique de Rangueil, 31077 Toulouse, France

F. Massazza Italcementi, Laboratorio Chimico Centrale, Bergamo, Italy

S. Mindess Department of Civil Engineering, University of British Columbia,
Vancouver, British Columbia, Canada

P.J.M. Monteiro Department of Civil Engineering, University of California, Berkeley,
CA 94720, USA

I. Odler Institute for Non-metallic Materials, Technical University Clausthal,
Germany

J.P. Ollivier Laboratoire Materiaux et Durabilite des Constructions, Complexe
scientifique de Rangueil, 31077 Toulouse, France

P.L. Pratt Department of Materials, Imperial College of Science and Technology,
London SW7 2BU, United Kingdom
*Professor Pratt died during the final stages of the preparation of this
Report. A biographical note appears on the following page.*

K.L. Scrivener Department of Materials, Imperial College of Science and Technology,
London SW7 2BU, United Kingdom

S.P. Shah NSF Center for Science and Technology of Advanced Cement-Based
Materials, Northwestern University, Evanston, Illinois, USA

H. Stang Department of Structural Engineering, Technical University of
Denmark, Lyngby, Denmark

Peter L. Pratt 1927–95

Peter Pratt graduated from Birmingham University in 1948. His post-graduate studies at the Cavendish Laboratory at Cambridge under Erwin Orowan and Sir Lawrence Bragg resulted in the award of his PhD (1952) on plasticity in sodium chloride. He worked at the Atomic Energy Research Establishment, Harwell and lectured at Birmingham University until 1958, when he was appointed Reader in Physical Metallurgy in the Department of Metallurgy at Imperial College London. He was Professor of Crystal Physics from 1963 to 1992 (Emeritus 1992). He contributed greatly to the deeper understanding of the development of microstructure during cement hydration, and its relation to materials properties and performance, both through his own work and through his leadership of the research group at Imperial College, which had an international reputation. Throughout his career he was awarded numerous prizes and honours.

Preface

With a view to taking stock of the subject and establishing areas of future investigation, the RILEM Technical Committee 'Interfaces in Cementitious Composites' set itself two objectives: to organize an International Conference at which the latest research results could be presented and to edit a state-of-the-art publication prepared by leading specialists. The first objective was achieved in October 1992, when an International Conference took place in Toulouse, the proceedings of which were published in *Interfaces in Cementitious Composites* by E & FN Spon. The second is achieved in the present book.

In the Introduction, Rector Jacques Farran, a pioneer in the subject, describes how the interfacial transition zone was discovered in the early 1950s, several thousand years after the first known concretes and about 200 years after the discovery – or rediscovery – by Apsdin and Vicat of the cements that are the origin of our modern concretes.

More and more scientists are becoming interested in the study of the interfacial transition zone between cement paste and aggregates. High performance concretes aside, in standard concretes this zone is the seat of primary defects and the preferred path for their propagation under mechanical, physical and chemical actions. The defects caused by such actions, working alone or in combination (the latter is more usual), can make a structure unfit to perform its function.

To understand this problem and also the conclusions that may be drawn from such knowledge, it is necessary to understand the physico-chemical mechanisms of cement hydration in the immediate vicinity of the aggregates. The microstructure of the interfacial transition zone in hardened concrete depends on these mechanisms. For this reason, the first two chapters of this book are devoted to their characterization and to the experiments used to study them.

The next two chapters are dedicated to the intrinsic properties of the interfacial transition zone: mechanical properties and transfer properties relating to the microstructure. Although many experiments, such as those for studying the microhardness of the transition zone, are now well established, our knowledge is still incomplete.

The final part of the book concerns the influence of the interfacial transition zone on the properties of a composite, i.e. of concrete. Although this zone is less than 15 – 20 microns thick, its influence on the response of concretes to external actions, and consequently on the durability of concrete structures, whether reinforced or not, is crucial.

Damage mechanisms under short-term mechanical actions are now understood from the phenomenological point of view, and although we still need to quantify them, studies are under way in various laboratories. However, study of the behaviour of the transition zone in response to long-term actions has only just begun.

Since they are interconnected in actual concretes, the interfacial transition zones clearly have an important influence on the transfer properties of concretes and, as a consequence, on their durability. Here again we are at the stage of initial studies and this subject will undergo significant development in the future, with very interesting results for practitioners.

The present work will be a reference not only for scientists, but also for practitioners, who will find inspiration from the formulations of concretes and from an improved understanding of the evolution of concrete structures in situ. Finally, the book will be a significant tool for both teachers and students.

Professor J.C. Maso
Toulouse, May 1994

Introduction: The transition zone – discovery and development

J. Farran
Paul Sabatier University, Toulouse, France

At the opening of the International RILEM Symposium on Interfaces in Cementitious Composites held in Toulouse in October 1992, the writer's introductory speech recalled the beginnings in Toulouse of a research programme that, even after more than 40 years, has not yet covered all aspects of the subject.

The story began in 1948. After the war, France was involved in the implementation of a vast rebuilding programme, particularly in the development of hydraulic energy. As part of this, many large dams, underground works and power stations were built. French engineers thus came into contact with rock in order to support structures on it, to go through it, or to use it in their concrete, in a way that nobody had done before. They soon discovered that, contrary to what they were taught or were supposed to understand from their physico-mathematical training, knowledge of the name and age of a rock did not give them any information on its behaviour in a technical sense (just as one cannot know an individual's reactions when one knows just his age and name).

What struck them particularly was the discovery that in the same gallery dug in the same very hard granite, which appeared to be stable and dry when it was built, the walls might crumble in some places after a few months, a few weeks or even a few days. At the same time, some of the largest extracted blocks exposed to the weather and stored outside for use in concrete for the lining, turned into an incoherent clay mass. But, in other passages, these changes did not take place although the rock was of exactly the same petrographic and crystallographic type and of the same chemical composition.

Apart from weathering, other extremely variable technical characteristics were noticed for rocks which nevertheless had the same petrography and mineralogy (for instance, drilling ability, 'crushability', capability to produce high quality aggregates).

Engineers were soon overwhelmed by the extent and economic consequences of the serious miscalculations linked to the behavioural variations of the rocks, which at first sight seemed incomprehensible because they were so unexpected. So, the engineers turned to the geologists. The geologists were accustomed to the study of large masses and were used to dealing with problems in terms of millions of years. They were not prepared to tackle these localised new types of problems and indeed they could not find any appropriate solution.

But the crystallographers, thanks to their small scale methodology and ability to study crystals up to the reticular lattice level, proved themselves able to detect quickly the beginnings of changes in the rocks, whereas the engineers noticed and suffered the consequences much later, and often too late to be able to find a solution.

Following requests from Electricité de France and other large public works companies, this writer was first prompted to study with some of his students, the main problems of rock weathering, its influencing factors, its measurement and rate, so as to define systematically what type of lining should be used for the underground structures.

In almost all cases, the concrete used for these linings was made with aggregates resulting

Interfacial Transition Zone in Concrete. Edited by J.C. Maso. RILEM Report 11.
Published in 1996 by E & FN Spon, 2–6 Boundary Row, London SE1 8HN. ISBN 0 419 20010 X.

quality was largely dependent on the state of the rock weathering. These observations, together with the study of the attachment of watertight cement-based coating cast against rock walls, logically brought us to examine the problem of adherence.

These two research areas, alteration and adherence, were obviously prompted by the technological demand without which they would probably not have been carried out. Indeed, both studies started on the building site but it must be emphasised that they have always been carried out with the thought and the means of the fundamental scientific research prevailing in our university laboratory.

The study of adherence carried out at the time of building the big hydroelectric works has proved to have a universal application since users of all types of concrete have become progressively more interested in this characteristic depending on their usage: air, maritime, nuclear etc.

The main published results are recalled here so as to make the link with the present work.

1 Microscopic observation of adherence on thin plates cut in mortars made from different pure mineralogical species, previously vacuum-impregnated by a coloured resin, clearly demonstrated the determined influence of the aggregate type (Farran, 1950).

2 During the setting and hardening of a defined cement in contact with a particular aggregate, some hydrated constituents settle in a specific manner in contact with the aggregate and the contact film type is different from that of the hydrated cement mass (Farran, 1953).

3 According to the cement and aggregate type, specific reactions can occur at their contact point, in a mortar or a concrete. Some of these reactions are destructive and others constructive, as a result of the formation of intermediate solid solutions or, even epitaxial attachments (Farran, 1956).

4 "For the values of the $c/(c+w)$ ratio corresponding to the practical field of composition of mortar and concrete the aggregates are surrounded by a particular paste, which constitutes a transition zone between them and the binder mass. Besides a composition difference, the texture of this zone is looser than that of the binder paste mass and its strength is lower. Under the action of external forces, fractures occur more often inside this zone" (Farran, 1956; Maso, 1967; Farran, Javelas, Maso and Perrin, 1972).

The interfacial zone, thus identified, seemed therefore to have a particularly important, even determining role, on several main characteristics of concrete (mechanical strength and frost resistance, for example). In fact, during recent years, many research works have been conducted on this subject and their authors gathered in Toulouse for the 1992 Symposium to discuss their points of view and to examine the current situation.

Besides numerous points of agreement, a divergence is to be noted related to the designation of this zone, that was originally called, after a long terminological debate, 'transition aureole' (Farran, Javelas, Maso and Perrin, 1972), putting aside, after careful consideration, expressions such as 'interfacial zone' or 'contact zone' that many people now seem to prefer. Lexicologists and etymologists in Greek and Latin would say that an aureole is golden (from the Latin aureus), plane and circular. It was long used by painters to encircle the head of Jupiter's descendants and then the Son of God's head; its usage was sparingly extended to the Saints only centuries later.

Bearing in mind this rather classical definition it is clear that the term is not really suitable for our field of activity - merely because it is not the most golden part of the concrete - far from it!

Nevertheless the usage became more widespread amongst scientists who called the 'aureole' the seat of any new formation arising from the meeting of two materials inducing a reaction when coming into contact. Metallurgists speak about an aureole in relation to welding and geologists have used the concept of 'aureole of metamorphism' to distinguish rocks of progressive variable types formed when an eruptive massif and a sedimentary massif meet. And here the word 'aureole' covers the four meanings of complete encircling, of interaction, thickness and gradient of composition, whereas the word 'zone' seems to the writer to convey a lot less than these four essential ideas.

But let us stop this analysis of terminology. The problem is neither insignificant, nor even marginal, but it is only semantics and changes nothing in the physico-chemical or mechanical reality. Nevertheless, having observed and described it in the first place in 1953, we wished to explain the decision, in 1972, to call the subject of our concerns the 'transition aureole'.

The main things which remain to be done are to broaden the knowledge, to assess the different characteristics, to study the properties and the consequences on the behaviour of cementitious composites, particularly during their technological applications.

Many researchers have worked and are still working on this subject, who participated in the International Symposium in 1992 to debate this question and to examine the state of the current knowledge. Considering the number and the quality of the participants, it was pleasing, but not surprising, to note that the research path that has been opened looks so promising.

On a personal note, I was glad to be able to tell the participants very sincerely that we were deeply touched that Toulouse had been selected to organize the Symposium and we were very grateful for their participation. The work on this topic has now reached another milestone with the completion of this RILEM Report.

References

Farran, J. (1950) *Bulletin de la Société d'Histoire Naturelle de Toulouse*, T.85-1950, p. 338.

Farran, J. (1953) *Comptes-rendus des séances de l'Académie des Sciences*, Paris T.237-1953, p. 73.

Farran, J. (1956) Thèse de Doctorat, Toulouse, in *Revue des Matériaux de Construction*, Paris, Nos 490, 491 and 492.

Maso, J.C. (1967) Thèse de Doctorat, Toulouse, in *Revue des Matériaux de Construction*, Paris, Nos 647, 648 and 649 (1969).

Farran, J., Javelas, R., Maso, J.C. and Perrin, B. (1972) *Comptes-rendus des séances de l'Académie des Sciences*, Paris, T.275-1972, p. 1467.

PART ONE

MICROSTRUCTURE

1

Characterisation of interfacial microstructure

Karen L. Scrivener and Peter L. Pratt

1.1 Introduction

It has long been appreciated that the microstructure of cement paste in the vicinity of an aggregate particle in mortar or concrete differs from that further away from the aggregate and from that in neat cement paste containing no aggregate. However, the interfacial region is an integral part of the whole microstructure so it is difficult to characterise the distinctive features of this region. Ideally the interfacial zone in real concrete should be studied, but only a limited number of techniques, mostly developed fairly recently, can be used to do this.

1.2 Specimen preparation

1.2.1 COMPOSITE SPECIMENS

In the 1950s, Farran (1956) developed a model composite specimen to study the interfacial zone specifically. This specimen, which has since been used extensively, consists of cement paste cast against a flat (usually polished) block of aggregate. The specimen can then either be split along the interface or cut perpendicular to it exposing the paste in the interfacial zone. The microstructure of the paste side of the interface may be characterised by X-ray diffraction (XRD) or surface analytical techniques as discussed below.

Despite the obvious advantages of this model interfacial specimen, it differs from real concrete in three important respects:

1. Only one piece of aggregate is present so the microstructure of the paste can adjust over an effectively infinite distance without encountering another aggregate particle as would be the case in concrete;
2. An extensive flat (and polished) surface will facilitate the formation of a film of water or more extensive bleeding at the interface;
3. The aggregate is not present during the mixing of the paste. The presence of aggregate during the mixing of concrete may have a significant effect on the paste by increasing local rates of shear and may allow the aggregate to acquire a coating of cement grains as it moves through the paste.

Later variants of this model composite specimen include the sandwich of paste between two

Interfacial Transition Zone in Concrete. Edited by J.C. Maso. RILEM Report 11.
Published in 1996 by E & FN Spon, 2-6 Boundary Row, London SE1 8HN. ISBN 0 419 20010 X.

flat pieces of aggregate used by Xie Ping *et al.* (1991) for measurement of electrical conductivity of the interfacial zone.

The third difference between the model and real concrete may be overcome, in part at least, by mixing the aggregate together with the paste, placing it in a mould and then pouring the paste onto it. Mitsui, Li and Shah (1991) used this technique with a small cylindrical piece of aggregate placed upright in a larger cylindrical mould. This composite specimen was used for push-out tests to characterise the mechanical properties of the interfacial zone.

1.2.2 REAL CONCRETE

If real concrete is to be studied there are three main possibilities for specimen preparation depending on the characterisation technique to be used.

Bulk comparisons
The pore size distribution of concretes and mortars as a whole may be compared with that of neat paste, by mercury porosimetry, magnetic resonance relaxation analysis (MRRA) or impedance spectroscopy. From these comparisons the effect of the interfacial zone on the overall pore size distribution may be determined but the spatial distribution of porosity in relation to the position of the paste–aggregate interface can only be inferred.

Fracture surfaces
A fracture path through concrete will pass through several interfacial regions, running parallel to the interface in some places and across it in others. Study of the fracture surface in the SEM allows the microstructure of these regions to be examined. This technique is very useful for qualitative characterisation, but, because the crack path favours weaker regions of the microstructure, fracture surfaces are unrepresentative of the concrete as a whole.

Cut and polished sections
Sections through bulk specimens prepared by mechanical cutting should be representative of the bulk if they are large enough. Cut and polished sections can be examined in the optical microscope (thin sections) or at higher resolution in the SEM with back scattered electrons. Despite their advantages over fracture surfaces, it is still not possible to determine the angle at which the plane of the section intersects the paste aggregate interface. Generally this angle of intersection will not be a right angle so the apparent width of the interfacial zone is expanded. It is also possible to prepare thin sections of the interface by ion beam milling for examination in the transmisision electron microscope (TEM), although with this technique only small regions can be examined.

1.3 Characterisation techniques

1.3.1 X-RAY DIFFRACTION (XRD)

X-ray diffraction is a valuable technique for studying the interfacial region in composite specimens. The technique developed by Grandet and Ollivier (1980a, 1980b) has been widely used to study the orientation of calcium hydroxide crystals in the interfacial zone. This technique entails splitting the composite specimen along the interface; an X-ray diffractometer trace of the cement paste side of the interface is then taken, a layer of paste is ground away and another XRD trace taken. In this way information is built up about calcium hydroxide and other crystalline components of the microstructure at increasing distances from the interface. The penetration of X-rays into cement paste will be of the order of 10 μm or more so there will be a degree of overlap between results from successive layers.

The most widespread application of this technique is to study the orientation of calcium hydroxide in the interfacial zone. For randomly oriented calcium hydroxide crystals the ratio of intensities for the peaks at 4.90 Å, from the {0001} basal planes, and at 2.628 Å, from the {$10\bar{1}1$} planes should be 0.74. Thus the ratio R = $(I_{\{1000\}}/I_{\{10\bar{1}1\}})/0.74$ is taken as a measure of the degree to which the CH crystals are preferentially oriented. Many studies have used this method to indicate that calcium hydroxide in the interfacial zone tends to be oriented with its basal plane parallel to the interface (c-axis perpendicular).

Recently the interpretation of results obtained with this technique has been questioned. It has been pointed out (Diamond, 1988) that only crystals oriented with the {0001} or {$10\bar{1}1$} planes parallel to the interface will contribute to the peaks measured by this technique. Detweiler *et al.* (1988) have studied the complete distribution of CH crystal orientations on pole figures and obtained results consistent with those obtained by the Grandet/Ollivier technique. Yuan *et al.* (1988) used rocking curves to study the degree of preferred orientation and obtained results qualitatively similar to those obtained with the Grandet/Ollivier technique.

Results obtained by Zürz and Odler (1987) indicate that the orientation index of calcium hydroxide changes with hydration time, water/cement ratio and other factors in a way that appears to be inconsistent with the usual interpretation of this index. In particular, the $(I_{\{1000\}}/I_{\{10\bar{1}1\}})$ peak intensity ratio was found to decline with progressing hydration, and preferred orientation was also detected by this method in bulk cement pastes. In a recent presentation, Maso (1991) suggested that the orientation index might be more correctly interpreted as a measure of the size of calcium hydroxide crystals in the interfacial zone, but this interpretation would not appear to explain the results obtained by Zürz and Odler. It is possible that calcium hydroxide crystals may tend to become oriented with their basal planes parallel to the interface during polishing, and thus that the orientation index is in part an artefact of the preparation process. Such reorientation would be more likely to occur when the microstructure was less dense (i.e. close to the interface, at high water/cement ratios and at shorter hydration times) and when the crystals were larger. Such an effect would, in part, explain the results of Zürz and Odler and the views of Maso and Grandet. However, when the cement paste was impregnated with resin before using the Grandet/Ollivier technique (in an attempt to avoid reorientation of the CH crystals during polishing) very similar results to the un-impregnated paste were obtained (Monteiro and Scrivener, 1991).

The progressive polishing technique has also been used to estimate the variation of weight fraction of calcium hydroxide and ettringite with depth (Monteiro and Mehta, 1985). Only semi-quantitative results can be obtained as the internal standard method cannot be used. However, it seems clear that the proportion of both CH and ettringite increase in the interfacial region.

1.3.2 SURFACE ANALYSIS (XPS, SIMS)

Composite specimens offer the possibility of characterising the microstructure of cement paste immediately adjacent to the aggregate with a range of surface analytical techniques.

X-ray photoelectron spectroscopy (XPS), otherwise known as electron spectroscopy for chemical analysis (ESCA), involves the use of soft X-rays under high vacuum to eject photoelectrons from atoms near the surface of solid materials. Typical depths of investigation are 5–10 nm and a chemical analysis of the surface layer can be obtained by measuring the kinetic energy of the photoelectrons. The technique has been used to measure C/S ratios in C_3S and cement powders which change rapidly during the first minutes of hydration.

Secondary ion mass spectrometry (SIMS) involves bombarding the surface with a beam of high energy ions, causing subsurface displacement cascades. Secondary ions emitted from the surface at the end of the cascade are analysed in a mass spectrometer. Low primary beam currents are necessary for good surface analysis. Modern instruments using a microfocused beam, rastered across the surface, can produce chemical maps with a lateral resolution of 0.2 μm and an analysis depth of only a few nanometres.

Very little work on mortars or concretes has been reported although the techniques could be used to study the composition of surface layers and near-surface layers of aggregates and of fibres either before mixing or after extraction from the hardened material.

1.3.3 BULK PORE SIZE DISTRIBUTION

Several techniques may be used to compare the pore structure in cement paste with that in concretes and mortars. Winslow and Liu (1990) compared the pore size distribution, measured using mercury intrusion porosimetry (MIP), of a plain cement paste with that same paste in mortar and in concrete. For mature specimens they found that the paste in mortar and concrete was more porous than the corresponding plain paste and the bulk of the additional pore volume, presumably in the interfacial zones, was in pore sizes larger than the threshold pore diameter for the plain paste. This is interesting but by itself difficult to interpret because the analysis is based upon an oversimplified geometrical model of pores as spheres or cylinders. A paper by Snyder *et al.* (1992) overcame this by using concepts from percolation theory to interpret the MIP measurements from mortars containing increasing sand contents. The pore size distribution, normalised per gram of cement, showed a dramatic change between 45% and 48% in the volume fraction of sand (Fig. 1.1). This, they concluded, indicated that the interfacial zones became linked together, or percolated, into a continuous pathway. They developed a continuum computer simulation model based on hard cores and soft shells, using some 10,000 aggregate particles. From this they concluded that the thickness of the interfacial zone which best explained the experimental MIP results was 15–20 μm.

Two other sophisticated physical techniques are available for the study of porosity in cement paste: magnetic resonance relaxation analysis (MRRA) and complex impedance spectroscopy. Unlike MIP, both must be used without drying the specimens because they rely on the presence of water in the pores; both are promising in the information they can give about the nature of the porosity. MRRA makes use of the difference in nuclear spin interactions between protons in bulk pore water and those encountering a liquid–solid interface. For a range of pore sizes inhomogeneously distributed over distances greater than a few μm, Halperin (1989) has shown that the inverse Laplace transform of the nuclear magnetic resonance (NMR) relaxation profile is simply related to the distribution of the local surface/volume ratios; this is equivalent to the local pore size distribution. He and co-workers used this technique to study porous rocks and ceramics and are now looking at cement paste.

The second technique, complex impedance spectroscopy, is valuable for studying changes

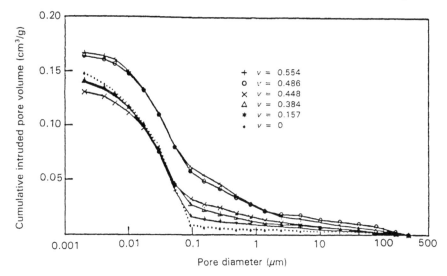

Fig. 1.1. Mercury intrusion porosimetry for mortars with varying sand contents (from Snyder *et al.*, 1992). Note the large difference in pore size distribution between mortars with sand contents of 0.448 and 0.486 by volume.

in a high conductivity phase, like pore water, embedded in a low conductivity matrix like the hydration products of cement. Mason and co-workers have applied it both to a range of ceramic materials and to the curing of cement pastes (Scuderi *et al.*, 1991; Christensen *et al.*, 1992). The important feature of this technique is that it enables the resistance of the bulk material to be separated from the electrode resistance unambiguously by measuring the impedance over a range of frequencies. The value of the real impedance at the minimum in the imaginary impedance between the electrode arc and the bulk arc gives the dc resistance of the bulk material. From this it is easy to obtain the dc conductivity of the bulk material. By expressing the pore fluid from the specimen in a press, the conductivity of the liquid phase can be measured separately. The difference between the conductivity of the free liquid and the liquid in the specimen gives a direct measure of the connectivity and tortuosity of the microstructure. Christensen *et al.* (1992) have reported systematic measurements of these conductivities for a cement paste, finding reasonable agreement with computer simulation. Beaudoin, at the Engineering Foundation Conference in Potosi in 1992, reported the results of applying the technique to study the development of the interfacial zone in mortars.

1.3.4 OPTICAL MICROSCOPE

Details of the spatial distribution of the porosity and of the phases in the hardened paste in the interfacial region can be seen with an optical microscope at a resolution limited by the wavelength of visible light, typically a few microns. Examination in transmitted polarised light of thin sections, prepared by cutting, grinding and polishing, is a well-known technique for concrete. Fluorescent resins are frequently used to fill the larger pores and in this way an indication of the water/cement ratio, the original mix design and the homogeneity of the porosity can be obtained.

1.3.5 ELECTRON MICROSCOPE

The transmission electron microscope (TEM) offers increased resolving power of the order of nanometres, because of the shorter wavelength of the beam of electrons, but the preparation of good thin foils, transparent to the electrons, is time consuming. For the highest resolution of the details of interfacial microstructure the TEM technique is necessary.

Scanning electron microscopy (SEM) has been used in two different modes of imaging: *secondary electron imaging* to study fracture surfaces where topographical features produce the contrast in the image and the large depth of focus is valuable, and *back-scattered electron imaging* to study flat polished surfaces where differences in back-scattering coefficient, dominated by atomic number differences, produce the contrast. Because back scattered electrons are generated from deeper in the specimen the resolution of the image is not as high as in the TEM, but the preparation of the specimen is much simpler. Detectors for energy-dispersive and wave-length dispersive X-ray analysis enable chemical compositions to be determined in regions excited by the electron beam. The scanning instruments produce a digitised image which can be stored on disc or transferred directly to image processing and image analysing systems. In this way gradients of microstructural composition can be determined to characterise the interfacial region (Scrivener and Gartner, 1988; Scrivener *et al.*, 1988a Scrivener *et al.*, 1988b).

1.4 Features of interfacial microstructure

There are two main components to the microstructure of the cement paste in the transition zone. First there is the thin layer of products which form directly on the aggregate surface, typically only a micron or so in thickness. This includes the products from any reaction there may be between the aggregate and the cement paste. In addition to this layer directly at the interface, there is the much larger region of paste in which the presence of the aggregate particle affects the original packing of the cement grains which, in turn, determines the subsequent development of microstructure. This region may extend some 50 μm or more out from the aggregate surface.

1.4.1 LAYER ON AGGREGATE SURFACE

The nature of the surface layer has been the subject of much debate. Barnes *et al.* (1978, 1979) used secondary electron imaging of fracture surfaces to study the layer which formed when cement paste was cast against a glass slide and subsequently the surface layer formed on aggregates in concretes. They characterised this as a 'duplex' layer of calcium hydroxide on the aggregate side and C-S-H on the cement paste side. The calcium hydroxide component is sometimes referred to as an epitaxial layer, but this is a misnomer. The term epitaxial implies a crystallographic orientation relationship between the product and substrate, but similar layers have been reported on glass and polythene which are non-crystalline (Yuan and Guo, 1987). However, any aggregate surface, whether in concrete or in a model specimen, provides an ideal heterogeneous nucleation site for the precipitation of either calcium hydroxide or C-S-H. Javels *et al.* (1974, 1975) and Zhang *et al.* (1988), using TEM of ion-thinned specimens from mortar, and Scrivener and Pratt (1986), using back-scattered electron (bse) imaging of polished surfaces, noted a similar thin layer formed directly on the aggregate surface but identified this as C-S-H. If a layer of CH does form part of the surface layer it is certainly well under a micron in thickness and so could not be detected by analysis of characteristic X-rays in the SEM, or by the XRD technique of Grandet and Ollivier. More recently Monteiro and Ostertag (1989) have reported the results of grazing angle X-ray scattering on the cement paste side of a composite specimen split at the interface, this gave a clear pattern for CH with no sign of any broad peak from C-S-H.

The presence of CH could, of course, depend on the nature of the specimen preparation and the size of the aggregate particles. Studies by Ping and Beaudoin (1991) of the electrical conductivity of concretes show a strong influence of the size of the aggregate on parameters related to the microstructure. A large, single piece of polished aggregate probably favours the formation of a continuous layer of water at the surface which would in turn favour the formation of CH. Whereas in mortars, used for most of the microscopic studies, large concentration gradients in the pore solution around aggregate particles would be less likely and consequently C-S-H would be most likely to precipitate on the aggregate surface.

1.4.2 PASTE AFFECTED BY PRESENCE OF AGGREGATE

The early work of Farran (1956) noted the lower density of the paste around the aggregate. Increased porosity around aggregate particles may also be observed by optical microscopy in thin sections of concrete impregnated with fluorescent dye, although a distinct interfacial zone is not obvious in good quality, low water/cement ratio concretes. Such qualitative observations have been confirmed by quantitative analysis of back-scattered electron images (Scrivener and Gartner, 1988; Scrivener *et al.*, 1988a, 1988b), which shows that the level of porosity steadily increases from a 'bulk' level some 20 μm from the interface to nearly twice this level in a 3 μm-wide band adjacent to the interface. This technique can also be used to determine the gradients in unreacted cement, calcium hydroxide and other hydration products. Typical plots from a one day old concrete are shown in Fig. 1.2. The gradient in porosity is reflected in the distribution of the unreacted cement which increases steadily from near zero adjacent to the interface to a 'bulk' level about 20 μm out. The amount of calcium hydroxide also increases to about twice its bulk value as the interface in approached.

1.4.2.1 Packing of cement particles at the interface
From the gradient of unreacted cement it is clear that the initial distribution of the cement grains around the aggregate particle is the principal factor in determining the subsequent microstructure. The aggregate is effectively a wall against which the cement grains must pack themselves. This effect is shown for an idealised 2-dimensional case with monosized circular particles in Fig. 1.3. Even though the particles are touching, the amount of empty space approaches 100% as the 'wall' is approached. In a 3-dimensional case of monosized spherical particles there may appear to be a 'film' of empty space at the interface even though particles are touching, Fig. 1.4. In concrete, the situation is further complicated by the fact that the cement grains have a wide range of sizes ranging from less than 1 μm up to 100 μm or so.

To look at the initial packing of cement particles in a concrete it would be necessary to examine a section of concrete before it is set, which is not practical. Escadeillas and Maso (1990) have attempted to study concretes as near to this initial state as possible by freeze-drying and resin-impregnating fresh concrete. This qualitative study indicated a zone of disturbance around the aggregate some 20 μm thick, in which more small particles and voids

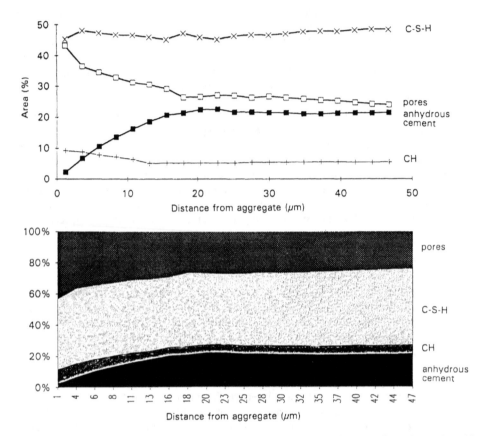

Fig. 1.2. Distribution of microstructural constitutuents in the interfacial region of a 1-day-old concrete.

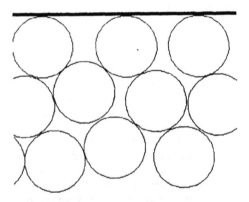

Fig. 1.3. Hypothetical packing of monosized circles against a 'wall' in two dimensions. Even though the circles are touching, the fraction of space occupied by the circles approaches zero as the interface is approached.

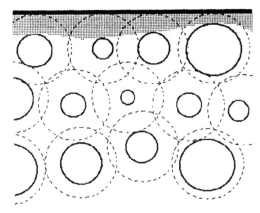

Fig. 1.4. Hypothetical two-dimensional section of monosized spheres packed against a 'wall' in three dimensions. Solid circles indicate profiles in plane of section while dotted lines indicate maximum cross-sections. Shaded area represents region at the interface which *appears* to be devoid of particles.

were observed. The initial distribution of grains in the interfacial region can also be deduced from relics and Hadley grains in a mature paste (Scrivener and Gartner, 1988). This indicates that small particles predominate close to the interface with larger particles lying further out.

A more quantitative approach is to calculate the initial distribution of anhydrous cement from the microstructural gradients measured in hardened pastes. This can be done from the analysis of concretes hydrated for 1 day, 28 days and 1 year by Crumbie and Scrivener (1993). To do this it has to be assumed that all the hydration products other than calcium hydroxide are deposited very close to the anhydrous cement from which they form. Although there is considerable evidence that C-S-H forms through solution (including the deposition of C-S-H on aggregate surfaces) this assumption is probably valid to a first approximation as the mobility of silicate species in solution is low. The initial distribution of anhydrous material so calculated is shown along with the equivalent original water/cement ratio in the interfacial region in Fig. 1.5.

If the aggregate particles are present during mixing the cement grains may become graded around the aggregate, with small particles packing close to the interface and larger particles tending to settle further away. This would be expected to reduce the gradients in porosity. Evidence of this grading can be deduced from the gradients of anhydrous (unreacted) cement measured after different amounts of hydration (Fig. 1.6). Even at 1 day the degree of reaction at the interface is higher than that in the bulk, indicating that the cement grains in this region are smaller and so have reacted more completely. The gradient measured in the concrete cured for 28 days suggests that the preponderance of small particles in the first 20 μm from the interface leads to a concentration of larger particles beyond this zone and that only beyond about 40–50 μm is the particle size distribution similar to that in the bulk. After curing for one year the gradient of anhydrous cement indicates that the effective width of the interfacial region is at least 50 μm confirming the long-range effect of the aggregate 'wall' on the packing of the cement grains.

Similar results to those obtained in real concretes were found by Garboczi and Bentz (1991) when the packing of cement grains around aggregate particles was simulated in a computer model.

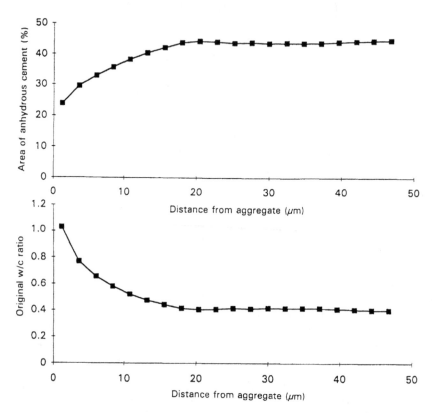

Fig. 1.5. Calculated initial distribution of anhydrous cement and effective water/cement ratio in the interfacial zone.

1.4.2.2 Effect of aggregate type

Grandet and Ollivier found that the degree of orientation of calcium hydroxide in the interfacial zone was affected by the mineralogical nature of the aggregate used. However, Crumbie (1994) found that the type of aggregate had relatively little effect on the microstructural gradients measured by image analysis in real concretes.

Comparison of as-received aggregates with polished aggregates indicates slightly narrower interfacial zones when the aggregate surface is rougher. Such differences probably occur as rough particles will have a greater tendency to acquire a 'coating' of small cement grains during mixing. In the case of some synthetic lightweight aggregates with very uneven surfaces this may result in quite significant amelioration of the microstructural gradients in the interfacial region (Ben Othman et al, 1988). Wu *et al.* (1988) have developed specially precoated aggregate particles and found significant improvement in mechanical properties when these are used in concrete.

1.4.2.3 Effect of mix design

Crumbie (1994) found that factors related to the mix design had some modifying effect on the packing of cement particles and hence on the microstructural gradients in

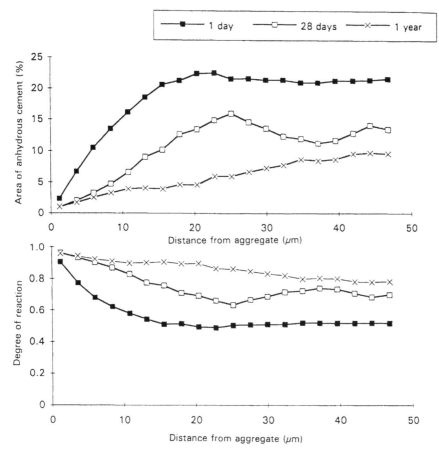

Fig. 1.6. Distribution of unreacted cement and calculated degree of reaction for concretes aged 1d, 28 d and 1 yr.

the interfacial zone. Low water/cement ratios and high aggregate/cement ratios appeared to promote efficient packing of cement particles at the aggregate and so minimised the width of the interfacial zone. Higher water/cement ratios and lower aggregate/cement ratios permitted rearrangement of the cement grains over greater distances resulting in wider interfacial zones in which microstructural gradients are shallower. However, the proportion of the microstructural constituents in the first 5 μm or so adjacent to the aggregate was relatively unaffected by mix design.

1.4.2.4 Differences in interfacial microstructure above and below aggregate
If significant settling occurs in the concrete before setting, the smaller cement particles tend to pile up on top of large aggregate particles (whose movement amongst the cement particles is impeded by their size). Similarly there will tend to be fewer cement particles below the aggregate. This will result in an effectively lower water/cement ratio above aggregate particles and an effectively higher water/cement ratio below. Such effects were reported by Hoshino (1988) and were noted to be more significant at higher overall water/cement ratios.

In the work of Crumbie (1994) there was relatively little difference in the microstructural gradients measured above and below aggregate particles at an overall water/cement ratio of 0.4. At the higher water/cement ratio of 0.6 the gradient in anhydrous cement was steeper above the aggregate particles and shallower below.

1.4.2.5 Redistribution of hydrates within the interfacial zone
If the amount of anhydrous material reacted and the amounts of CH and C-S-H formed at different distances from the aggregate are compared, the redistribution of hydrates within this zone can be investigated. Such a comparison involves several assumptions and will be subject to errors. However, Fig. 1.7 shows some results from the work of Crumbie (1994) in which the general trends are thought to be valid. This figure shows amounts of CH and C-S-H calculated as being in excess of the amounts expected if all the hydration products were precipitated in the immediate vicinity of the anhydrous cement from which they formed. It is apparent that there is a fairly large-scale redistribution of calcium hydroxide. Due to the greater availability of space close to the interface the growth of crystals in this region is relatively unrestricted. Calcium and hydroxide ions diffuse from paste further away from the aggregate leading to a relative excess in the first 10 μm adjacent to the interface and a relative deficit in the next 40 μm or so. Significant redistribution has occurred already after

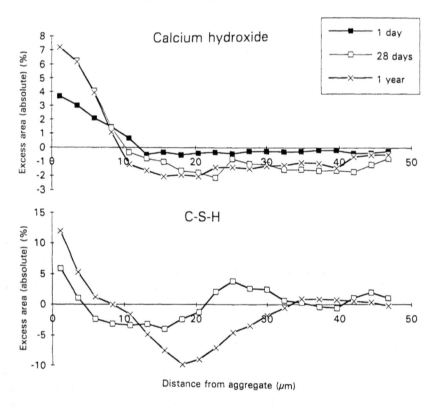

Fig. 1.7. Calculated 'excess' deposition of products, above those formed from the cement originally present, in the interfacial zone.

1 day of hydration and has considerably increased by 28 days. Further curing from 28 days to 1 year results in little further change presumably owing to the reduced diffusion rates of ions through a more compact structure. Very roughly speaking, half the CH in a 3 μm band adjacent to the interface in a mature paste has formed from the reaction of anhydrous phases outside this region.

The redistribution of C-S-H is less marked due to the lower mobility of silicate species in solution. The calculation assumed that there was negligible redistribution in the one day paste although this assumption is probably not entirely valid. After 28 days there is a slight excess of C-S-H in the first 5 μm nearest to the interface and this excess increases somewhat after curing for 1 year. However, in a mature paste only about 20% or so of C-S-H in a 3 μm band adjacent to the interface has formed from the reaction of material outside this region.

The net effect of this redistribution of hydrates is to ameliorate the gradient in porosity initially set up by the packing of the anhydrous particles. However, the interfacial zone continues to have a higher porosity than the bulk. This is mainly the result of the initial packing but is also contributed to by the phenomenon of one-sided growth noted by Garboczi and Bentz (1991). In the bulk paste, pore space may become filled with hydrates by the migration of species into that space from all directions, and on the scale of several microns there is no redistribution of hydrates. Close to the interface, however, hydrates must migrate from the cement paste side. This phenomenon results in an increased porosity in the first 5 μm or so adjacent to the aggregate, above and beyond that arising from the packing effect.

1.4.2.6 'Width' of interfacial zone

The packing of unreacted cement grains around aggregate particles results in a region of affected microstructure of a width of similar order of size to the largest cement particles. However, the width of the zone in which the porosity is noticeably different from that of bulk paste is probably less than 20 μm (a median size for cement grains). Transport properties depend critically on the degree of connectivity of the pore structure, which evidence from MIP and impedance measurements suggests is most affected within this narrower zone.

When very fine particles are present, such as silica fume or even inert particles such as carbon black, these pack preferentially close to the aggregate and there is little or no increase in porosity in the interfacial region (Scrivener *et al.*, 1988a).

1.5 Summary

- In fresh pastes a thin layer of hydration products precipitates on aggregate particles. In compact and well mixed systems such as good quality, low water/cement ratio concrete this layer is essentially C-S-H. In composite specimens or at higher water/cement ratios, where bleeding may occur, calcium hydroxide may precipitate on the aggregate surface, especially on the lower surface.
- Beyond this surface layer the interfacial microstructure is essentially determined by the packing of cement grains against the aggregate 'wall'.
- Packing is largely determined by the particle size distribution of the cement paste. However, the water/cement ratio, aggregate/cement ratio, mixing and settling of the concrete play a role.
- This packing of the cement grains results in microstructural gradients extending some 50 μm from the interface into the cement paste in which:

- the amount of anhydrous cement decreases as the interface is approached;
- the amount of porosity increases;
- the amount of CH increases and may become oriented;
- the amount of ettringite increases.

• In mature pastes the major change in porosity is confined to 15–20 μm around the aggregate, reflecting the effects of 'one-sided' growth of hydrates as well as packing.
• In this thinner interfacial zone the connectivity of the pore structure appears to increase as well as the amount of pore space. As a consequence this region is of significance in determining the transport properties and durability of concrete.

1.6 References

Barnes, B.D., Diamond, S. and Dolch, W.L. (1978) The contact zone between Portland cement paste and glass "aggregate" surfaces, *Cement and Concrete Research*, Vol. 8, pp. 233–44.

Barnes, B.D., Diamond, S. and Dolch, W.L. (1979) Micromorphology of the interfacial zone around aggregates in Portland cement mortar, *Journal of the American Ceramic Society*, Vol. 62, pp. 21–4.

Ben-Othman, B., Scrivener, K.L. and Buenfeld, N.R. (1988) Permeability and microstructure of lightweight aggregate concrete, presented at Institute of Ceramics, Annual Convention, Durham, April.

Christensen, B.J., Mason, T.O., Jennings, H.M., Bentz, D.P. and Garboczi, E.J. (1992) Experimental and computer simulation results for the electrical conductivity of Portland cement paste, *Advanced Cementitious Systems: Mechanisms and Properties*, Materials Research Society, Vol. 245, pp. 259–64.

Crumbie, A.K. (1994) PhD Thesis, University of London.

Detweiler, R.J., Monteiro, P.J.M., Wenk, H.-R. and Zhong, Z. (1988) Texture of calcium hydroxide near the cement paste–aggregate interface, *Cement and Concrete Research*, Vol. 18, pp. 823–9.

Diamond, S. (1988) personal communication, cited in Detweiler *et al.* (1988).

Escadeillas, G.C. and Maso, J.C. (1990) Approach of the initial state in cement paste, mortar and concrete, *Advances in Cementitious Materials*, (ed. S. Mindess), American Ceramics Society, Ceramic Transactions, Vol. 16, pp 169–84.

Farran, J. (1956) Contribution mineralogique a l'etude de l'adherence entre les constituants hydrates des ciments et les materiaux enrobes, *Revue des Materiaux de Construction*, Nos 490-491, pp. 155-7, No. 492, pp. 191–209.

Garboczi, E.J. and Bentz, D.P. (1991) Digital simulation of the aggregate–cement paste interfacial zone in concrete, *Journal of Materials Research*, Vol. 6, No. 1, pp. 196-201.

Grandet, J. and Ollivier, J.-P. (1980a) Nouvelle methode d'etude des interface ciments-granulats, *Proceedings of the 7th International Congress on the Chemistry of Cement*, Editions Septima, Paris, Vol. III, pp. VII. 85-9

Grandet, J. and Ollivier, J.-P. (1980b) Orientation des hydrates au contact des granulats, *Proceedings of the 7th International Congress on the Chemistry of Cement*, Editions Septima, Paris, Vol. III, pp. VII. 63-8.

Halperin, W.P., D'Orazio, F., Battacharaja, S. and Tarczon, J.C. (1989) Magnetic resonance relaxation analysis of porous media, *Molecular Dynamics in Restricted Geometries*, (eds J. Klafter and J.M. Drake), John Wiley and Sons Inc., chap. 11, p. 311.

Hoshino, M. (1988) Difference of the w/c ratio, porosity and microscopical aspect between the upper boundary and the lower boundary of the aggregate in concrete, *Materials and Structures*, Vol. 21, pp. 336–40.

Javels, R., Maso, J.C. and Ollivier, J.-P. (1974) Realisation de lames ultra-mince de mortier pour observation directe au microscope electronique par transmission, *Cement and Concrete Research*, Vol. 4, pp. 167–76.

Javels, R., Maso, J.C., Ollivier, J.-P. and Thenoz, B. (1975) Observation directe au microscope electronique par transmission de la liason pate de ciment-granulat dans des mortier de calcite et de quartz, *Cement and Concrete Research*, Vol. 5, pp. 285–94.

Maso, J.C. (1991) presentation at 3rd European Conference on Microscopy of Building Materials, Barcelona, September.

Mitsui, K. Li, Z. and Shah, S.P. (1991) A new method of characterising properties of the paste–aggregate zone, Presented at Materials Research Society Symposium, Advanced Cementitious Systems: Mechanisms and Properties.

Monteiro, P.J.M and Mehta, P.K. (1985) Ettringite formation at the aggregate–cement interface, *Cement and Concrete Research*, Vol. 15, pp. 378–80.

Monteiro, P.J.M. and Ostertag, C.P. (1989) Analysis of the aggregate–cement paste interface using grazing incidence X-ray scattering, *Cement and Concrete Research*, Vol. 19, pp. 987–8.

Monteiro, P.J.M. and Scrivener, K.L. (1991) unpublished work.

Ping, X., Beaudoin, J.J. and Brousseau, R. (1991) Flat aggregate–Portland cement paste interfaces: I Electrical conductivity models, *Cement and Concrete Research*, Vol. 21, No. 4, pp. 515–22.

Ping, X. and Beaudoin, J.J. (1991) Effect of aggregate size on the transition zone properties at the Portland cement paste interface, *Cement and Concrete Research*, Vol. 21, No. 6, pp. 999–1005.

Scrivener, K.L. and Pratt, P.L. (1986) A preliminary study of the microstructure of the cement paste–aggregate bond in mortars, *Proceedings of the 8th International Congress on the Chemistry of Cement*, Rio de Janeiro, Vol. III, pp. 466–71.

Scrivener, K.L. and Gartner, E.M. (1988) Microstructural gradients in cement paste around aggregate particles, *Bonding in Cementitious Composites*, (eds S. Mindess and S.P. Shah), Materials Research Society, Vol. 114, pp. 77–86.

Scrivener, K.L., Bentur, A. and Pratt, P.L. (1988a) Quantitative characterisation of the transition zone in high strength concretes, *Advances in Cement Research*, Vol. 1, pp. 230–9.

Scrivener, K.L., Crumbie, A.K. and Pratt, P.L. (1988b) A study of the interfacial region between cement paste and aggregate in concrete, *Bonding in Cementitious Composites*, (eds S. Mindess and S.P. Shah), Materials Research Society, Vol. 114, pp. 87–8.

Scuderi, C.A., Mason, T.O. and Jennings, H.M. (1991) Impedance spectra of hydrating cement pastes, *Journal of Materials Science*, Vol. 26, pp. 349–53.

Snyder, K.A. Winslow, D.N., Bentz, D.P. and Garboczi, E.J. (1992) Effects of interfacial zone percolation on cement based composite transport properties, *Advanced Cementitious Systems: Mechanisms and Properties*, (eds F.P. Glasser *et al.*), Materials Research Society, Vol. 245, pp. 265–70.

Winslow, D.N. and Liu, D. (1990) The pore structure of paste in concrete, *Cement and Concrete Research*, Vol. 20, pp. 227–35.

Wu, X., Li, D., Wu, X. and Minshu, T. (1988) Modification of interfacial zone between aggregate and cement paste, *Bonding in Cementitious Composites*, (eds S. Mindess and S.P. Shah), Materials Research Society, Vol. 114, pp. 35–40.

Yuan C.Z. and Guo, W.J. (1987) Bond between marble and cement paste, *Cement and Concrete Research*, Vol. 17, pp. 544–52.

Yuan, C.Z., Min, Z.W. and Jiang, L. (1988) Determination of preferred orientation degree of portlandite by using rocking curve of diffraction line, *Materials and Structures*, Vol. 21, pp. 329–35.

Zhang, X., Groves, G.W. and Rodger, S.A. (1988) The microstructure of cement aggregate interfaces, *Bonding in Cementitious Composites*, (eds S. Mindess and S.P. Shah), Materials Research Society, Vol. 114, pp. 89–95.

Zürz, A. and Odler, I. (1987) XRD studies of Portlandite present in hydrated Portland cement paste, *Advances in Cement Research*, Vol. 1, No. 1, pp. 27–30.

2

Development and nature of interfacial microstructure

A. Bentur and I. Odler

2.1 Introduction

The nature of the microstructure of the interfacial zone has been studied extensively. The structure of the interfacial zone in the vicinity of the inclusion in the paste is often quite different from that of the bulk paste away from the inclusion surface. The development of this special structure is, to a large extent, dependent on the nature of the fresh mix, in particular with respect to the formation of water-filled spaces around the inclusion's surface (Escadeillas and Maso, 1991). The nature and density of the hydration products which develop in this zone will depend on its initial porosity, the composition of the matrix and the composition of the inclusion's surface. The geometry of the inclusion, and the nature of the process by which the cementitious composite is produced, may play a major role in controlling the interfacial microstructure. Much of this influence may be due to the effect of these parameters on the distribution of water-filled spaces and unhydrated cement grains in the vicinity of the surface, in the fresh state of the composite.

The present chapter provides an overview of the various parameters which may influence the process by which the interfacial microstructure is formed, and its nature in the mature composite. This topic is best addressed by classifying the field of cementitious composites into two types: cementitious composites with particulate filler (i.e. aggregates in concretes), and fibre-reinforced cements.

2.2 Concretes

2.2.1 GENERAL CHARACTERISTICS

The formation of the interfacial zone in concretes is due, to a large extent, to the formation of water-filled spaces around the aggregates in the fresh mix (Diamond, 1986; Bentur and Cohen 1987; Scrivener, Bentur and Pratt, 1988), i.e. the water distribution in the paste surrounding the aggregate is non-uniform, with effective larger water/cement ratio in the vicinity of the aggregate surface (Hoshino, 1989). This effect may be due to bleeding, as well as to a wall effect which prevents effective filling of the space adjacent to the aggregate with cement grains having a size of 10 μm or more. As a result, the space around the aggregates is less effectively filled by hydration products, and at the same time there is greater tendency for CH (Ca(OH)$_2$) and ettringite to develop in this space, since these two

Interfacial Transition Zone in Concrete. Edited by J.C. Maso. RILEM Report 11.
Published in 1996 by E & FN Spon, 2–6 Boundary Row, London SE1 8HN. ISBN 0 419 20010 X.

compounds form and deposit preferentially in large pores. A schematic description of the process by which this zone is formed is shown in Figs 2.1(a) and (b). An SEM micrograph demonstrating this microstructure is shown in Fig. 2.2.

Since the formation of the interfacial zone is influenced by bleeding effects, it is to be expected that its structure around the aggregate will not be uniform, with the highest porosity interfacial zone forming below the aggregate, and the lowest above it. This has been confirmed in quantitative studies of the microstructure (Hoshino, 1989), and is shown schematically in Fig. 2.3. The width of this zone can be estimated by various means, showing it to be in the range of 10 to 100 μm. This is also the same order of magnitude of size as the cement grains, thus consistent with the hypothesis that the formation of this zone is affected by a wall effect leading to inefficient packing of the cement grains in the vicinity of the surface.

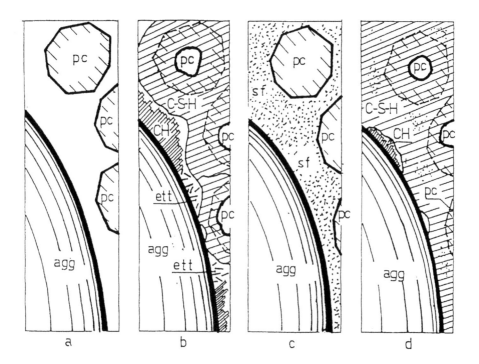

Fig. 2.1 Schematic description of the mode and nature of formation of the interfacial zone around aggregates in a cementitious mix of Portland cement only (a,b) and Portland cement plus silica fume (c,d) (after Goldman and Bentur, 1989): (a) Fresh concrete without silica fume, showing the water-filled space around the aggregate surface, due to bleeding and inefficient cement grain packing at the boundary; (b) The interfacial zone of the mature system in (a) showing filling of the interfacial zone with CH and C-S-H, and the remnants of porous pockets and zones, some filled with needle-like material; (c) Fresh concrete with silica fume, showing the silica fume particles filling the space around the aggregate which was occupied by water in the concrete without silica fume in (a); (d) The less porous interfacial zone in the mature system of (c). PC - Portland cement grains; SF - silica fume particles; CH - Ca(OH)$_2$; C-S-H - calcium silicate hydrate gel; ETT - ettringite.

Fig. 2.2 Interfacial microstructure around an aggregate particle in a cement paste, showing voids and large deposits of CH (after Bentur and Cohen, 1987).

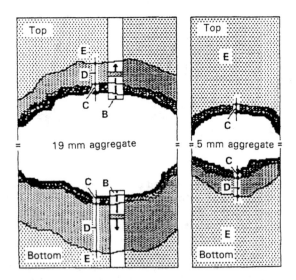

Fig. 2.3 Schematic description of the interfacial zone, showing its width to be greater beneath the aggregate and lowest above it.

Quantitative analysis of the interfacial microstructure by the Toulouse method of orientation index (Maso, 1980) or by BSE image analysis (Bentur and Pratt, 1988) of pores, suggest that this zone may be described as one in which there is a gradient in microstructure, starting with high porosity and highly oriented CH crystals in the proximity of the surface, with gradual change in these characteristics as one moves away from the surface. Such a gradient is shown in Fig. 2.4 (control curve, for Portland cement concrete).

These general characteristics of the interfacial zone may be affected by various parameters, either in the paste matrix itself or in the composition of the aggregate. Some of them are outlined in the following sections.

2.2.2 EFFECT OF CHEMICAL ADMIXTURES

Very little has been published on the effect of chemical admixtures on the structure of the interfacial zone. Ollivier, Grandet and Hanna (1988) investigated the effect of an added naphthalene-sulphonate superplasticizer on such structures (Portland cement with and without 10% added silica fume, carbonate rock, 4% naphthalene sulphonate). They found that the orientation of CH crystals, within the interfacial zone as well as its thickness, were not altered distinctly by the additive.

2.2.3 EFFECT OF MINERAL ADMIXTURES

From among the mineral admixtures, the effect of silica fume on the structure of the interfacial zone has been studied most extensively. It has been reported that the thickness of

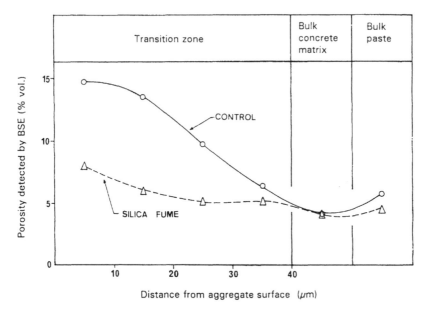

Fig. 2.4 Gradients in porosity in the interfacial zone adjacent to the aggregate in Portland cement concrete (control curve) and in silica fume concrete (after Scrivener *et al.*, 1988).

the interfacial zone becomes reduced as silica fume is added to the system (Monteiro and Mehta, 1986a). The amount of CH within the interfacial zone becomes significantly reduced if compared to systems that do not contain silica fume, especially after longer hydration times (Bentur and Cohen, 1987; Wang Jia *et al.*, 1986; Chen Zhi Yuan and Zhang Xio-Zhong, 1986; Aitcin, Sarkar and Diatka, 1987; Odler and Zürz, 1988). At the same time, the preferential orientation of CH within the interfacial zone, as defined by the orientation index, is reduced as well (Ollivier, Grandet and Hanna, 1988; Monteiro and Mehta, 1986a; Wang Jia *et al.*, 1986). Thus a more dense and homogenous structure, similar to that of the bulk paste, develops under these conditions (Diamond, 1986; Bentur and Cohen, 1987; Scrivener *et al.*, 1988; Goldman and Bentur, 1989; Labri and Bijen, 1990), as observed by SEM analysis, qualitatively and quantitatively (Fig. 2.4). It has also been reported, however, that the addition of silica fume results in coarser pores close to the interface (Feldman, 1986). These changes in the structure of the interfacial zone are accompanied by an improved cement paste–aggregate bond and thus an increase of the interfacial cleavage energy (Wang Jia *et al.*, 1986; Chen Zhi Yuan and Zhang Xio-Zhong, 1986; Aitcin, Sarkar and Diatka, 1987; Odler and Zürz, 1988). An increase of the effective fracture energy was also reported.

These characteristics of the interfacial zone, induced by the addition of silica fume can be accounted for by its physical nature, which allows a more efficient packing at the surface and less bleeding in the fresh mix, as well as the chemical nature of the silica fume which is highly pozzolanic and reacts to form C-S-H, at the expense of CH. This is shown schematically in Figs 2.1(c) and (d).

No significant changes were found in the structure of the interfacial zone if type F fly ash (5%) was added to Portland cement (Odler and Zürz, 1988).

2.2.4 EFFECT OF CEMENT COMPOSITION

A comparison of different types of Portland cement, i.e. 'ordinary', 'white' and 'sulphate-resistant', revealed only minor differences in the structure of the interfacial zone (Odler and Zürz, 1988). In samples made with the iron-rich 'sulphate-resistant' cement, the ettringite crystals exhibited a different, less acicular morphology than seen in the OPC and 'white' cement samples. The former cement also exhibited better bonding characteristics than the latter two.

Data on the structure and characteristics of the interfacial zone in the presence of cements other than Portland cements are scarce. Frigione, Marchese and Sersale (1986) reported cement–aggregate bond strengths with high slag cements to be distinctly higher than found with Portland cement. Monteiro and Mehta (1985) reported data on the structure of the interfacial zone formed in the presence of type K expansive cement. Such interfacial zones were characterized by a high concentration of ettringite. The preferential orientation of CH crystals was reduced. At the same time the bond strength increased.

2.2.5 EFFECT OF AGGREGATE COMPOSITION, TREATMENT AND TYPE

The chemical composition of normal weight aggregate has only a small influence on the nature of the interfacial zone, and this may be considered as a second order effect. The composition of the interfacial zone can be modified when the aggregate reacts somewhat with the matrix (e.g. calcite and dolomite); carboaluminates or calcium carbonate–calcium hydroxide complex may deposit in the interfacial zone, and the aggregate surface may become somewhat etched. Such influences of the nature of the aggregates have been discussed in several publications (Odler and Zürz, 1988; Chen and Wang, 1987; Monteiro and Mehta, 1986b).

The nature of the microstructure of the interfacial zone suggests the presence of two weak links: one at the actual interface, and the second in the pores in the interfacial zone, somewhat away from the actual interface. It was demonstrated by Odler and Zürz (1988) that the weak link depends on the aggregate type and matrix. Bond tests indicated that with basalt aggregate the failure occurred at the actual interface, while with quartzite, limestone, marble and greywacke it took place within the interfacial zone (Fig. 2.5). With the incorporation of silica fume, the location of the debonded zone was moved further into the interfacial zone. The cleavage bond strength increases as the location of debonding moved deeper into the interfacial zone (Odler and Zürz, 1988).

These observations imply that when no chemical interaction with aggregate takes place, the weak link is the actual interface; when this interface is somewhat strengthened by chemical interaction, the weak link becomes the more porous area in the interfacial zone. This implies that in order to improve bond in these systems there is a need to modify the whole microstructure of the interfacial zone. Attempts to enhance bond strength by polymer treatment of the actual interface were not successful (Popovics, 1987) and this was attributed to the fact that debonding took place within the bulk of the transition zone, away from the actual interface.

Therefore, treatments with mineral solutions or with silica fume, which apparently interact with greater volume of the cement paste matrix in the vicinity of the aggregate surface, are effective in enhancing bond.

Treatments with solutions were reported by Zimbelmann (1987) and Wu Xuequan (1986). Zimbelmann demonstrated the effectiveness of surface treatments in reducing the thickness of the interfacial zone (use of surface active agents to reduce the thickness of the water film in the fresh mix), or in promoting reactions which improve physical and chemical bond by

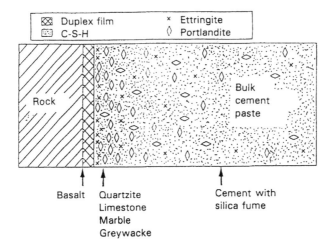

Fig. 2.5 Debonding zones in systems of different aggregates and binder composition (after Odler and Zürz, 1988).

reacting mainly with the CH of the interfacial zone (solutions of water glass and pozzolans, for example). Such treatments can lead to a dramatic improvement in bond strength. At the age of one year some of the treated aggregate exhibited bond strength greater than the paste strength (Zimbelmann, 1987). Wu Xuequan *et al.* (1986) also evaluated treatments with various reagents, showing that the improved bond could lead to increase in compressive and flexural strength by about 30 to 40%. The treatment is simple (mixing the aggregates with the treating solutions prior to adding other ingredients) and may not be too difficult to apply.

Around lightweight aggregates the microstructure of the interfacial zone can be quite different. It has been observed that in these systems the interfacial zone can be much denser than in normal weight aggregate, and this can be attributed to the absorption of free water that accumulates around the aggregate (Min-Hong Zhang and Gjorv, 1989 and 1990). For such an influence to take place the aggregate must be partially dried. If the aggregates possess some pozzolanic reactivity, additional changes in the interfacial microstructure might be expected.

2.3 Fibre-reinforced cements

2.3.1 INTRODUCTION

Fibre-reinforced cements and concretes have become viable construction materials. A wide variety of properties can be obtained from these composites, depending on the production process, type and geometry of fibres, and the composition of the cementitious matrix. In such systems, the properties are much more sensitive to processes taking place at the interface, compared to concretes. The reasons are that the fibre surface area is considerably larger than that of aggregates in concrete, and the more important role that the fibres play in enhancing the mechanical properties of the matrix (strength and toughness). Therefore, much more attention has been given to the study of interfaces in fibre-reinforced cement systems. The object of the present section is to review the various interfacial microstructures in such composites and highlight the wide variety of morphologies that may be obtained, depending on composition and production variables.

2.3.2 STEEL FIBRES

The microstructure of the interfacial zone in steel fibres has been studied by several investigators (Al Khalaf and Page, 1979; Page, 1982; Pinchin and Tabor, 1978; Bentur, Diamond and Mindess, 1985; Bentur, Diamond and Mindess, 1985; Bentur, Gray and Mindess, 1986) and it may be generally concluded that it is quite similar to that observed around aggregates in concrete. A schematic presentation is provided in Fig. 2.6. The interfacial zone is rich in CH, which is mostly in direct contact with the fibre surface. It is also quite porous, making it different from the microstructure of the bulk paste. Some of these characteristics can be clearly observed in the SEM micrograph in Fig. 2.7. The layer of CH surrounding the fibre is not necessarily uniform (Fig. 2.8), and occasionally a duplex film is seen around the fibre (Fig. 2.9).

The mode of formation of this interfacial zone is apparently similar to that which takes place around aggregate inclusions in concrete: a water-filled space is formed around the fibre in the fresh mix and CH precipitates from solution in this void, with the fibre surface acting

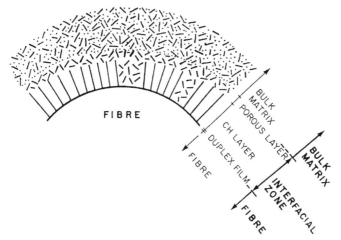

FIBRE

BULK MATRIX

POROUS LAYER

CH LAYER

DUPLEX FILM

FIBRE

BULK MATRIX

INTERFACIAL ZONE

FIBRE

Fig. 2.6 Schematic presentation of the microstructure of the interfacial zone around steel fibres (after Bentur *et al.*, 1986).

5 μm

Fig. 2.7 SEM micrograph showing the CH layer and the higher porosity material in the interfacial zone around a steel fibre (after Bentur *et al.*, 1985).

Fig. 2.8 SEM micrograph showing the discontinuity in the CH layer engulfing a steel fibre in the interfacial zone (after Bentur *et al.*, 1985).

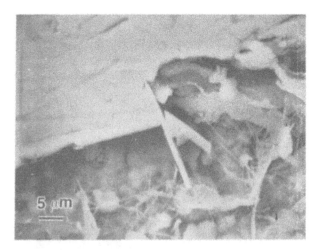

Fig. 2.9 SEM micrograph showing remnants of a thin duplex film which formed around the surface of a steel fibre (after Bentur *et al.*, 1985).

as a nucleating site. The rest of the space is filled with C-S-H, and occasionally with some ettringite; since the void volume is large in this zone, the filling is not efficient, and considerable porosity still remains. The 'pockets' in the discontinuous layer of CH (Fig. 2.8) tend to be filled up to a greater extent with needle-like material, and it is in this zone that the duplex film can be observed.

2.3.3 ASBESTOS FIBRES

Microstructural analysis of the interfacial zone in asbestos cement was carried out in conjunction with fracture studies of asbestos-cement composites. Akers and Garrett (1983a, 1983b) and May (1979) showed that, in the actual composite, a considerable portion of the fibres was in the form of bundles (Fig. 2.10). The diameter of the bundles ranged from 1 to 300 μm. Occasionally, some very fine fibrils can be seen at the edge of such a bundle, and they are interlocked with hydration products (Fig. 2.11).

In general, however, many of the fibrils which form the original asbestos fibre bundle (Fig. 2.12), are dispersed in the cementitious matrix. The microstructure around them is quite dense, and no large porosity or CH crystals can be observed in the interfacial zone, in contrast to the cement–steel interface discussed previously. It was also found that the C/S ratio of the hydrated material in the bulk, and at the interfacial zone, was quite similar. This dense and uniform microstructure at the interfacial zone may be related to the more hydrophillic surface of the asbestos fibres, compared to other reinforcing inclusions, and to the processing of the asbestos–cement composites by the Hatschek machines, which probably eliminates the tendency for formation of water-filled spaces in the vicinity of the fibres in the fresh product.

Fig. 2.10 The bundled nature of the asbestos fibres in a cementitious matrix, as seen in in-situ testing (after Akers and Garrett, 1983a).

Fig. 2.11 Fibrils at the edge of asbestos bundle, interlocked with hydration products (after Akers and Garrett, 1983a).

Fig. 2.12 Asbestos fibre bundle composed of numerous tiny fibrils, prior to being mixed and dispersed in the cementitious matrix (after Akers and Garrett, 1983b).

Opoczky and Pentek (1975) studied the microstructural changes which may take place at the interface during ageing. They concluded that the fibres can undergo partial carbonation and, in addition, cement hydration products which crystallize on the surface of the asbestos fibres and along the cleavage planes between fibrils, may react chemically with the fibres, especially in the presence of carbon dioxide. Majumdar (1975) also hypothesized growth of hydration products between the fibrils during ageing.

2.3.4 GLASS FIBRES

Characteristic features, which set the interfacial microstructure in glass fibre–cement composite, are the geometry of the glass fibre and the production process for the composite. The glass fibres are produced in the form of rovings, with each roving being made up of several strands (Fig. 2.13). Each roving is made up of about 200 filaments of approximately 10 μm diameter (Fig. 2.13). The most common process for production of glass fibre cement (GRC) composite is by the spray technique, which is efficient in dispersing the roving into individual strands (Fig. 2.14a), but the filaments in each strand remain grouped together (Fig. 14b). As the spaces between the filaments are rather small, less than about 3 μm, cement grains, which are usually much bigger, cannot penetrate between them and, as a result, very few hydration products grow between the filaments during the first few weeks of curing (Stucke and Majumdar, 1976; Bentur, 1986a) (Fig. 2.15).

The space between the filaments may gradually become filled with hydration products after prolonged ageing. Since there are no cement grains between the filaments, the development of hydration products is apparently by a through solution mechanism, with the mode of growth being dependent, to a large degree, on the chemistry of the cementitious system and

Fig. 2.13 Geometry of glass fibres: roving made up several strands, with each strand composed of approximately 200 filaments.

Fig. 2.14 Dispersion of glass fibre in the cement matrix as obtained by the spray production technique: (a) Strands dispersed in the matrix; (b) Individual filaments remain grouped together in the strand.

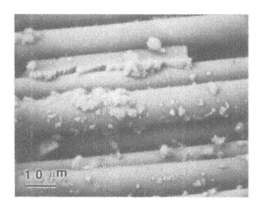

Fig. 2.15 Microstructure of 28-day-old GRC composite showing very little growth of hydration products between the filaments (after Bentur, 1986a).

the nature of the surface of the glass fibre, as the surface may serve as a nucleation site. It was shown that, in the case of the first generation of alkali-resistant (AR) glass fibres (e.g. CemFIL-1), it was CH that mainly deposited in this zone, forming a dense and highly crystalline microstructure after several years of natural ageing or after a few weeks of accelerated ageing (Stucke and Majumdar, 1976; Bentur, 1986a, 1986b; Mills, 1981; Bentur, Ben-Bassat and Schneider, 1985; Bijen and Jacobs, 1982) (Fig. 2.16).

As it was shown that the massive growth of CH around glass filaments was responsible, at least in part, for the embrittlement of GRC (Stucke and Majumdar, 1976; Bentur, 1986b; Bentur, Ben-Bassat and Schneider, 1985), attempts were made to modify the interfacial microstructure around the filaments by modifying the surface of the fibres or the composition of the cementitious matrix. The second generation of AR glass fibres (e.g CemFIL-2), was obtained by surface treatment of the earlier AR fibres (Proctor, Oakley and Litherland, 1982), and the resulting durability improvement was attributed to the formation of less dense interfacial microstructure after ageing (Fig. 2.17). Similar modification in the interfacial microstructure could be achieved by blending the Portland cement matrix with slag and pozzolans, to reduce or eliminate the formation of CH at the interface and thus improve the durability performance (Bentur, 1986a; Proctor, Oakley and Litherland, 1982; Majumdar and Singh, 1982; Leonard and Bentur, 1984; Singh and Majumdar, 1981; Singh, Majumdar and Ali, 1984; Singh and Majumdar, 1985). The use of non-Portland cement matrices, such as high alumina cement (Majumdar and Singh, 1982), supersulphated cement (Majumdar, Singh and Evans, 1981; Singh and Majumdar, 1987) and the recently developed Japanese CGC cement (Hayashi, Sato and Fujii, 1985; Tanaka and Uchida, 1985; Akihama *et al.*, 1987) (calcium silicate-C_4A_3S-CS-slag) was also found to be effective in eliminating CH formation at the interface, and thus leading to improved durability performance. It should be noted that, in some of these composites, durability problems may arise due to the limited long-term performance of the matrix itself (e.g. high alumina cement, supersulphated cement).

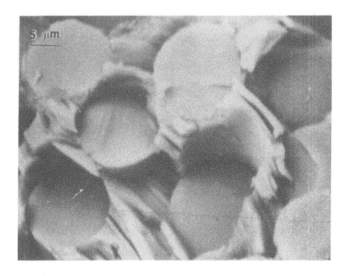

Fig. 2.16 The spaces between filaments in an aged AR GRC composite (CemFIL-1), showing growth of dense and massive CH crystals (after Bentur, 1986a).

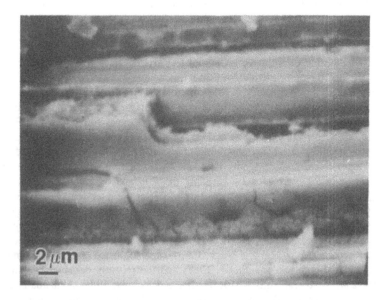

Fig. 2.17 The spaces between filaments in an aged AR GRC composite (CemFIL-2), showing growth of porous, amorphous material, apparently C-S-H (after Bentur, 1986a).

An alternative approach, based on the treatment of the glass fibres to eliminate growth of CH at the interface, was to immerse the strands in a silica fume slurry prior to the production of the composite, so that the tiny silica fume particles that were deposited in the spaces between the filaments (Fig. 2.18) eliminated the growth of CH at the interface and let to a composite of improved durability performance (Bentur and Diamond, 1987; Bentur, 1989a).

Several studies led to the use of polymer latex-modified cement as a means for controlling the interfacial microstructure and the durability performance (Singh, Majumdar and Ali, 1984; West, De Vekey and Majumdar, 1985; West, Majumdar and De Vekey, 1986; Majumdar, Singh and West, 1987; Bijen, 1980; Bijen, 1983; Jacobs, 1986a; Jacobs, 1986b; Jacobs and Bijen, 1985). The tiny polymer particles (smaller than approximately 0.1 μm) penetrate between the spaces in the glass filaments (Bijen, 1980), and, after initial curing, the coalesced polymer phase between the filaments prevents growth of the rigid hydration products in these spaces. This mechanism was proved useful for commercial production of GRC composites of improved durability.

It should be emphasized that the discussion in the previous paragraphs referred to alkali-resistant glass, where the chemical attack of the glass by the alkaline cementitious matrix is eliminated, or its rate is considerably reduced. Thus, the control of the properties of composites with such glass, in particular long-term performance, is based mainly on means for modifying the growth of hydration products in the vicinity of the glass surface. However, if E-glass fibres are being used, the main interfacial effect of practical significance is of a chemical reaction which causes corrosion of the glass (Larner, Speakman and Majumdar, 1976), and it shows up first of all by the formation of large defects on the glass filament surface (Fig. 2.19).

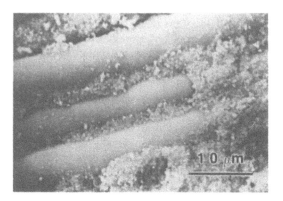

Fig. 2.18 Impregnation of the spaces between glass filament with silica fume from slurry, to prevent formation of CH at the interface (after Bentur, 1989a).

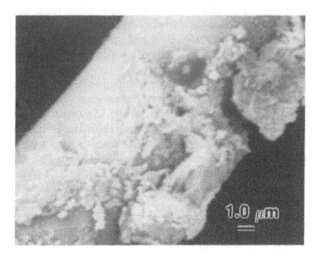

Fig. 2.19 The surface of E-glass fibre removed from an aged composite with a Portland cement matrix (after Bentur, 1986a).

2.3.6 CELLULOSE FIBRES

Cellulose fibres, produced by the pulp process, are increasingly used as a partial or full replacement of asbestos in asbestos-cement composites manufactured by the Hatschek process (Studinka, 1986; Bleiman, Bulens and Robin, 1984). The pulping process separates the wood species into the individual cellulose fibre (cell), which is a fibre having the typical structure resembling a hollow tube with a diameter of approximately 10 μm. In the production process of the composite the individual cellulose fibres are uniformly dispersed in the matrix.

The interfacial microstructure in cellulose-cement composites, produced by the Hatschek process or by a similar technique, has been studied with a view to understanding the relations between bonding and failure mode of the composite (brittle vs ductile) at early age, and possible interfacial microstructural changes which occur over time, that may account for embrittlement on ageing, as is observed in some cases.

In the unaged composite, whether cured normally or in an autoclave, the interfacial microstructure does not contain any massive CH, as observed in steel fibres or aged glass fibres. However, there seems to be a difference between a normally cured and autoclaved cured composite: the interfacial microstructure is more porous in the former (Fig. 2.20), and much more dense in the latter (Bentur and Akers, 1989a, 1989b) (Fig. 2.21). This difference in microstructure may account for the difference in the mode of failure of the two composites which is by fibre pull-out in the normally cured composite (Fig. 2.20) and fibre fracture in the autoclaved cured composite (Fig. 2.21).

Cellulose-cement composites may change their properties over time. It was suggested that densification of the interfacial microstructure over time can transform the failure mode from pull-out to fibre fracture, thus leading to loss of toughness (Bentur and Akers, 1989; Davies,

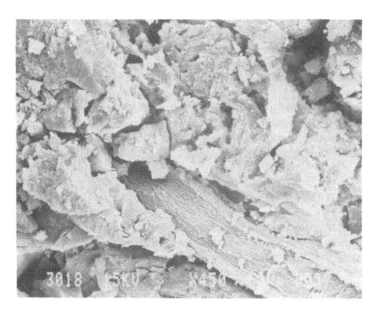

Fig. 2.20 Interfacial microstructure around a cellulose fibre in a composite cured normally (after Bentur and Akers, 1989a).

Fig. 2.21 Interfacial microstructure around a cellulose fibre in a composite cured in an autoclave (after Bentur and Akers, 1989b).

Campbell and Coutts, 1981). Such a densification in a normally cured composite can be seen in Fig. 2.22 (compare Fig. 2.22 with Fig. 2.20). During ageing in an environment containing carbon dioxide, as may happen in natural weathering, additional effects, on top of the densening of the interfacial microstructure may occur, and they include petrification of the fibre (Bentur and Akers, 1989). This shows in filling up of the core of the cellulose fibre with hydrated material (Fig. 2.23), as well as deposition of some calcium silicate in the cell wall itself (Pirie *et al.*, 1990). These changes may lead to effective strengthening of the fibre itself, as well as to improved fibre–matrix bond, and may account for the increase in strength and reduction in toughness after the ageing in CO_2 environment of the normally cured composite (Akers and Studinka, 1989).

2.3.7 SYNTHETIC FIBRES

Synthetic fibres used for cement and concrete reinforcement, cover a range of compositions, from low modulus polypropylene to high modulus fibres such as carbon, aramid, acrylic and polyvinyl alcohol (PVA) fibres. These fibres are different also in their geometry: the polypropylene fibres which are usually applied in cement and concrete mixes are of the fibrillated film type, whereas the high modulus fibres are usually single filaments with diameters less than 10 μm.

The fibrillated polypropylene film can be mixed with concrete, and during the mixing operation the film network is separated into individual fibre units (Fig. 2.24) that can be 10 to 100 μm thick and 100 to 600 μm wide (Rice, Vondran and Kunbargi, 1988; Bentur, Mindess and Vondran, 1989). The fibrillated film can also be used to produce thin sheet material by impregnating it with a cement slurry using special production techniques

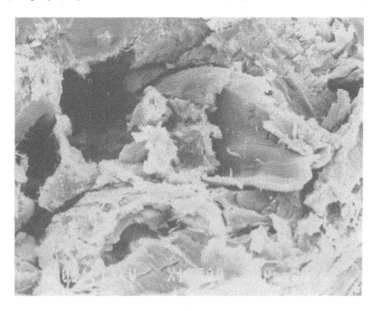

Fig. 2.22 Interfacial microstructure around a cellulose fibre in a composite cured normally, after ageing in CO_2-free environment (after Bentur and Akers, 1989a).

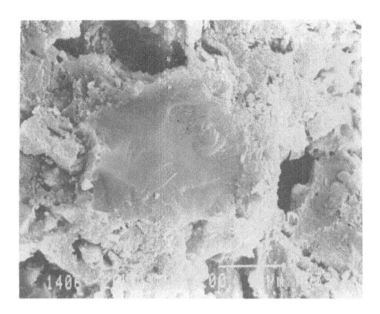

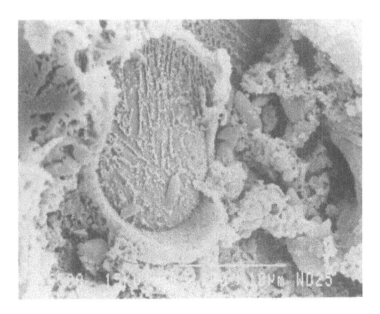

Fig. 2.23 Interfacial microstructure around and within a cellulose fibre in a composite cured normally, after ageing in CO_2 environment (after Bentur and Akers, 1989a): (a) Petrified fibre with the core filled with hydrated material and a dense, surrounding microstructure; (b) Petrified fibre after slight etching in HCl, revealing the fibre cell wall and the products deposited in the core.

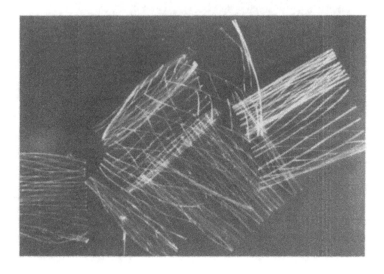

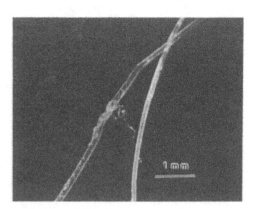

Fig. 2.24 Fibrillated polypropylene film before mixing in concrete (a) and its separation to individual units after mixing in concrete (b) (after Bentur *et al.*, 1989).

(Hannant and Zonsveld, 1980). The interfacial microstructure observed in commercial fibrillated polypropylene fibres in a concrete mix is quite dense and uniform, showing no tendency for formation of CH at the interface as may occur with other fibres (Fig. 2.25). It was suggested that the formation of this dense interface was apparently due to surface treatments provided during the production of the fibres (Bentur, Mindess and Vondran, 1989). A study of the effects of various surface treatments of polypropylene in monofilament form suggested that the more effective treatments are the ones which involve wetting agents and physical roughening of the surface. This shows up by the formation of a denser interfacial microstructure, as well as improved performance of the composite (Peled, Guttman and Bentur, **Accepted for publication,** *Cement and Concrete Composites*).

The high modulus monofilament synthetic fibres (carbon, aramid, acrylic and PVA) have been reported to provide a highly effective reinforcing effect when mixed in a cementitious matrix (Akiyama *et al.,* 1986: Hikasa *et al.,* 1986: Hahne *et al.,* 1987; Odler, 1988; Katz and Bentur, 1991; Bentur, 1989b, 1990). This is probably due to their effective dispersion in the matrix, which in combination with the high surface area of the thin filaments provide a high bond. In the few microstructural studies reported in such composites, it was shown that the interfacial microstructure was uniform, without deposition of CH. The interface observed in composites with a low water/cement ratio matrix produced by intensive mixing was very dense, as seen, for example, for carbon fibres in Fig. 2.26. It was reported that increase in strength and reduction in toughness in PVA composites (Hikasa *et al.,* 1986) might be due to improved bond, which may be associated with densening of the interface. On the other hand, in the case of acrylic fibres dispersed in a concrete matrix, the interfacial

Fig. 2.25 Dense and uniform interfacial microstructure around a polypropylene fibre in a concrete matrix (after Bentur *et al.,* 1989).

Fig. 2.26 Interfacial microstructure around a carbon filament in a low water/cement ratio matrix (after Katz and Bentur, 1991).

microstructure tended to be more porous (Odler, 1988) than that observed for the carbon fibres in Fig. 2.26. Such differences may be due to the nature of the fibre surface and the mixing process. However, in both cases the interfacial microstructure was uniform without any CH, in contrast to steel and glass fibres. This difference may have to do with the uniform dispersion of the fine filaments, which perhaps do not entrap bleeding water around them, and thus enable a more uniform microstructure to be developed at the interface.

2.4 Conclusions

The present chapter provided an overview of the development of the interfacial zone in cementitious composites, and the various parameters which may affect its microstructure.

In the case of concretes, the interfacial microstructure is not very sensitive to the composition of the Portland cement and the composition of the aggregate. Drastic differences in the interfacial microstructure are observed only in special cases, for example, when silica fume is added to the matrix or when lightweight aggregate is being used.

The review of the interfacial microstructure of the different fibre-cement composites, indicated the existence of a wide variety of morphologies, which depend on the nature of the fibre itself (composition and geometry), the composition of the matrix and the production process. In addition, considerable changes may take place at the interface in some of the systems, due to ageing effects. In view of these, it is not possible to provide a general model or description which will account for all the morphologies described here. Each system should be addressed separately, taking into consideration all the factors mentioned above.

It should be noted, however, that the interfacial microstructure may play an important role in controlling the properties of fibre-cement composites, in particular since the fibre specific surface area is much larger than other inclusions that are incorporated in cementitious matrices, like aggregates. Therefore, resolving the nature of the interfacial microstructure and understanding of the processes by which it was generated may provide valuable tools to control the properties of the composite, early in its lifetime, as well as after ageing. Some of these aspects are reviewed in chapter 9.

2.5 References

Aitcin, P.C., Sarkar, S.L. and Diatka, Y. (1987) Microstructural study of different types of very high strength concretes, *Microstructural Development During the Hydration of Cement*, (eds L. Struble and P.W. Brown), Materials Research Society, Vol. 85, pp. 261–72.

Akers, SA.S. and Garrett, G.G. (1983a) Fibre-matrix interface effects in asbestos-cement composites, *Journal of Materials Science*, Vol. 18, pp. 2200–8.

Akers, SA.S. and Garrett, G.G. (1983b) Observations and predictions of fracture in asbestos-cement composite, *Journal of Materials Science*, Vol. 18, pp. 2209–14.

Akers, SA.S. and Studinka, J.B. (1989) Ageing behaviour of cellulose fibre cement composites in natural weathering and accelerated tests, *International Journal of Cement Composites and Lightweight Concrete*, Vol. 11, pp. 93–7.

Akihama, S., Suenaga, T. and Banno, T. (1986) Mechanical properties of carbon fibre reinforced cement composites, *International Journal of Cement Composites and Lightweight Concrete*, Vol. 8, pp. 21–34.

Akihama, S., Suenaga, T., Tanaka, M. and Hayashi, M. (1987) Properties of GRC with low alkaline cement, *Fiber Reinforced Concrete, Properties and Applications*, (eds S.P. Shah and G.B. Batson), ACI SP-105, American Concrete Institute, Detroit, pp. 289–9.

Al Khalaf, M.N. and Page, C.L. (1979) Steel mortar interfaces: microstructural features and mode of failure, *Cement and Concrete Research*, Vol. 9, pp. 197–208.

Bentur, A. (1986a) Microstructure and performance of glass fibre-cement composites, *Research on the Manufacture and Use of Cements*, (ed. G. Frohnsdorff), Engineering Foundation, pp. 197–208.

Bentur, A. (1986b) Mechanisms of potential embrittlement and strength loss of glass fibre reinforced cement composites, *Proceedings of Durability of Glass Fiber Reinforced Concrete Symposium*, (ed. S. Diamond), Prestressed Concrete Institute, Chicago, pp. 109–23.

Bentur, A. (1989a) Silica fume treatments as a means for improving the durability of glass fibre reinforced cements, *ASCE Journal of Materials in Civil Engineering*, Vol. 1, pp. 167–83.

Bentur, A. (1989b) The role of the interface in controlling the performance of high quality cement composites, *Advances in Cement Manufacture and Use*, (ed. E. Gartner), Engineering Foundation, pp. 227–37.

Bentur, A. (1990) Microstructure, interfacial effects and micromechanics of cementitious composites, *Advances in Cementitious Materials*, (ed. S. Mindess), The American Ceramic Society, pp. 523–49.

Bentur, A. and Akers, SA.S. (1989a) The microstructure and ageing of cellulose fibre reinforced cement composites cured in a normal environment, *International Journal of Cement Composites and Lightweight Concrete*, Vol. 11, pp. 99–109.

Bentur, A. and Akers, SA.S. (1989b) The microstructure and ageing of cellulose fibre reinforced autoclaved cement composites, *International Journal of Cement Composites and Lightweight Concrete*, Vol. 11, pp. 111–15.

Bentur, A., Ben-Bassat, M. and Schneider, D. (1985) Durability of glass fibre reinforced cements with different alkali resistant glass fibres, *Journal of the American Ceramic Society*, Vol. 68, pp. 203–8.

Bentur A. and Cohen, M.D. (1987) The effect of condensed silica fume on the microstructure of the interfacial zone in Portland cement mortar, *Journal of the American Ceramic Society*, Vol. 70, pp. 738–43.

Bentur, A. and Diamond, S. (1987) Direct incorporation of silica fume into strands as a means for developing GFRC composites of improved durability, *International Journal of Cement Composites and Lightweight Concrete*, Vol. 9, pp. 127–36

Bentur, A., Diamond, S. and Mindess, S. (1985a) The microstructure of the steel fiber-cement interface, *Journal of Materials Science*, Vol. 20, pp. 3610–20.

Bentur, A., Diamond, S. and Mindess, S. (1985b) Cracking processes in steel fiber reinforced cement paste, *Cement and Concrete Research*, Vol. 15, pp. 331–42.

Bentur, A., Gray, R.J. and Mindess, S. (1986) Cracking and pull-out processes in fibre reinforced cementitious materials, *Developments in Fibre Reinforced Cement and Concrete*, (eds R.N. Swamy, R L. Wagstaffe and D.R. Oakley), Proceedings RILEM Conference, Sheffield, paper 6.2.

Bentur, A., Mindess, S. and Vondran, G. (1989) Bonding in polypropylene fibre reinforced concrete, *International Journal of Cement Composites and Lightweight Concrete*, Vol. 11, pp. 153–8.

Bijen, J. (1980) Glass fibre reinforced cement: improvements by polymer addition, *Advances in Cement Matrix Composites*, (eds D.M. Roy, A.J. Majumdar, S.P. Shah and J.A. Manson), Materials Research Society, pp. 239–49.

Bijen, J. (1983) Durability of some glass fibre reinforced cement composites, *Journal of the American Concrete Institute, Proceedings*, Vol. 80, pp. 305–11.

Bijen, J. and Jacobs, M. (1982) Properties of glass fibre reinforced polymer modified cement, *Materials and Structures*, Vol. 15, pp. 445–52.

Bleiman, C., Bulens, M. and Robin, P. (1984) Alternatives for substituting asbestos in fibre cement products, *Proceedings Conference on High Performance Roofing Systems*, Plastic and Rubber Institute, London, pp. 8.1–8.12.

Chen, Z.Y. and Wang, J.G. (1987) Bond between marble and cement paste, *Cement and Concrete Research*, Vol. 17, pp. 544–52.

Chen, Z.Y. and Zhang Xio-Zhong (1986) Distribution of $Ca(OH)_2$ and the C-S-H phase in oolitic marble/hydrated cement paste, *Proceedings of the 8th International Congress on the Chemistry of Cement*, Rio de Janeiro, Vol. 3, pp. 449–53

Davies, G.W., Campbell, M.D. and Coutts, R.S.P. (1981) A SEM study of wood fibre reinforced cement composite, *Holzforschung*, Vol. 35, pp. 201–4.

Diamond, S. (1986) The microstructure of cement paste in concrete, *Proceedings of the 8th International Congress on the Chemistry of Cement*, Rio de Janeiro, Vol. I, pp. 122–47.

Escadeillas, G.C. and Maso, J.C. (1991) Approach of the initial state in cement paste, mortar and concrete, *Advances in Cementitious Materials*, (ed. S. Mindess), American Ceramic Society, pp. 169–84.

Feldman, R.F. (1986) The effect of sand/cement ratio and silica fume on the microstructure of mortar, *Cement and Concrete Research*, Vol. 16, pp. 31–9.

Frigione, G., Marchese, B. and Sersale, R. (1986) Microcracking propagation in flexural loaded Portland and high slag cement concretes, *Proceedings of the 8th International Congress on the Chemistry of Cement*, Rio de Janeiro, Vol. 3, pp. 478–84.

Goldman A. and Bentur, A. (1989) Bond effects in high strength silica fume concretes, *ACI Materials Journal*, Vol. 86, No. 5, pp. 440–7.

Hahne, H., Karl, S. and Worner, J.D. (1987) Properties of polyacrylonitrile fibre reinforced concrete, *Fibre Reinforced Concrete Properties and Applications*, (eds S.P. Shah and G.B. Batson), ACI SP-105, American Concrete Institute, Detroit, pp. 211–23.

Hannant, D.J. and Zonsveld, J.J. (1980) Polyolefin fibrous networks in cement matrices for low cost sheeting, *Philosophical Transactions of the Royal Society of London*, A294, pp. 591–7.

Hayashi, M., Sato, S. and Fujii, H. (1985) Some ways to improve durability of GFRC, *Proceedings of Durability of Glass Fibre Reinforced Concrete Symposium*, (ed. S. Diamond), Prestressed Concrete Institute, Chicago, pp. 270–84.

Hikasa, J. Genba, T. Mizobe, A. and Okazaki, M. (1986) Replacement for asbestos in reinforced cement products - "Kuralon" PVA fibres, properties and structure, *International Man-Made Fibres Congress*, Dornbirm, The Austrian Chemical Society.

Hoshino, M. (1989) Relationship between bleeding, coarse aggregate and specimen height of concrete, *ACI Materials Journal*, Vol. 86, pp. 125–90.

Jacobs, M.J.N. (1986a) Durability of PGRC, design implications, Part 1, *Betonwerk und Fertigteil-Technik*, Vol. 52, pp. 228–33.

Jacobs, M.J.N. (1986b) Durability of PGRC, design implications, Part 2. *Betonwerk und Fertigteil-Technik*, Vol. 52, pp. 756–61.

Jacobs, M.J.N. and Bijen, J. (1985) Durability of Forton polymer modified GFRC, *Proceedings of Durability of Glass Fiber Reinforced Concrete Symposium*, (ed. S. Diamond), Prestressed Concrete Institute, Chicago, pp. 1293–1304.

Katz, A. and Bentur, A. (1991) High performance fibres in high strength cementitious matrices, *High Performance Fiber Reinforced Cement Composites*, (eds H.W. Reinhardt and A.E. Naaman), RILEM Workshop, Mainz, E & FN Spon, London, pp. 139–48.

Larbi, J. and Bijen, J.M. (1990) Orientation of calcium hydroxide at the Portland cement paste–aggregate interface in mortars in the presence of silica fume, *Cement and Concrete Research*, Vol. 20, pp. 461–70.

Larner, L.J., Speakman, K. and Majumdar, A.J. (1976) Chemical interactions between glass fibres and cement, *Journal of Non Crystalline Solids*, Vol. 20, pp. 43–74.

Leonard, S. and Bentur, A. (1984) Improvement of the durability of glass fiber reinforced cement using blended cement matrix, *Cement and Concrete Research*, Vol. 14, pp. 717–28.

Mai, Y.W. (1979) Strength and fracture properties of asbestos-cement mortar composites, *Journal of Materials Science*, Vol. 14, pp. 2091–102.

Majumdar, A.J. (1975) Properties of fibre cement composites, *Fibre Reinforced Cement and Concrete*, (ed. A.M. Neville), Proceedings RILEM Symposium, The Construction Press, UK, pp. 279–313.

Majumdar, A.J. and Singh, B. (1982) Non Portland cement GRC. Information Paper IP 7/82, Building Research Establishment, UK.

Majumdar, A.J., Singh, B. and Ali, M.A. (1981) Properties of high-alumina cement reinforced with alkali resistant glass fibres, *Journal of Materials Science*, Vol. 16, pp. 2597–607.

Majumdar, A.J., Singh, B. and Evans, T.J. (1981) Glass fibre-reinforced supersulphated cement, *Composites*, Vol. 12, pp. 171–83.

Majumdar, A.J., Singh, B. and West, J.M. (1987) Properties of GRC modified by styrene-butadiene rubber latex, *Composites*, Vol. 18, pp. 61–4.

Maso, J.C. (1980) The bond between aggregates and hydrated cement paste, *Proceedings of the 7th International Congress on the Chemistry of Cement*, Paris, Editions Septima, Vol. I., pp. VII-1/3 – VIII-1/15.

Mills, R H. (1981) Preferential precipitation of calcium hydroxide on alkali resistant glass fibres, *Cement and Concrete Research*, Vol. 11, pp. 689–98.

Min-Hong Zhang and Gjorv, O.E. (1989) Backscattered electron imaging studies on the interfacial zone between high strength lightweight aggregate and cement paste, *Advances in Cement Research*, Vol. 2, pp. 141–6.

Min-Hong Zhang and Gjorv, O.E. (1990) Microstructure of the interfacial zone between lightweight aggregate and cement paste, *Cement and Concrete Research*, Vol. 20, pp. 610–8.

Monteiro, P.J.M. and Mehta, P.K, (1985) Ettringite formation at the aggregate cement paste interface, *Cement and Concrete Research*, Vol. 15, pp. 378–80.

Monteiro, P.J.M. and Mehta, P.K. (1986a) Improvement of the aggregate–cement paste transition zone by grain-refinement of hydration products, *Proceedings of the 8th International Congress on the Chemistry of Cement*, Rio de Janeiro, Vol. 3, pp. 433–7.

Monteiro, P.J.M. and Mehta, P.K. (1986b) The transition zone between aggregate and type K expansive cement, *Cement and Concrete Research*, Vol. 16, pp. 127–34.

Odler, I. (1988) Structure and mechanical properties of Portland cement-polyacrylnitril fibre composites, *Bonding in Cementitious Composites*, (eds S. Mindess and S.P. Shah), Materials Research Society, Proceedings, Vol. 114, pp. 153–8.

Odler, I. and Zürz, A. (1988) Structure and bond strength of cement-aggregate interfaces, *Bonding in Cementitious Composites*, (eds S. Mindess and S.P. Shah), Materials Research Society, Proceedings, Vol. 114, pp. 21–7.

Ollivier, J.P., Grandet J. and Hanna, B. (1988) Effect of superplasticizer and reactive condensed silica fume on the mortar–coarse aggregate bond (in French), *Proceedings of the 8th International Congress on the Chemistry of Cement*, Rio de Janeiro, Vol. 4, pp. 204–9.

Opoczky, L. and Pentek, L. (1975) Investigation of the 'corrosion' of asbestos fibres in asbestos cement sheets weathered for long times, *Fibre Reinforced Cement and Concrete,* (ed. A.M. Neville), Proceedings RILEM Symposium, The Construction Press, UK, pp. 269–76.

Page, C.L. (1982) Microstructural features of interfaces in fibre cement composites, *Composites*, Vol. 13, pp. 140–4.

Peled, A., Guttman, H. and Bentur, A. Treatments of polypropylene fibres as a means to optimize for the reinforcing efficiency in cementitious composites. Accepted for publication, *Cement and Concrete Composites.*

Pinchin, D.J. and Tabor, D. (1978) Interfacial phenomena in steel fibre reinforced cement I. Structure and strength of the interfacial region, *Cement and Concrete Research*, Vol. 8, pp. 15–24.

Pirie, B.J., Glasser, F.K., Schmidt-Henco, C. and Akers, SA.S. (1990) Durability studies and characterization of the matrix and fibre-cement interface of asbestos free fibre-cement products, *Cement and Concrete Composites*, Vol. 12, pp. 233–44.

Popovics, S. (1987) Attempts to improve the bond between cement paste and aggregate, *Materials and Structures,* Vol. 20, pp. 32–8.

Proctor, B.A., Oakley, D.R. and Litherland, K.L. (1982) Developments in the assessments and performance of GRC over 10 years, *Composites*, Vol. 13, pp. 173–9.

Rice, E.K, Vondran, G.L. and Kunbargi, H.O. (1988) Bonding of fibrillated polypropylene fibre to cementitious materials, *Bonding in Cementitious Composites,* (eds S. Mindess and S.P. Shah), Materials Research Society, Proceedings, Vol. 114, pp. 145–52.

Scrivener, K.L., Bentur, A. and Pratt, P.L. (1988) Quantitative characterization of the transition zone in high strength concretes, *Advances in Cement Research*, Vol. 1, No. 4, pp. 230–7.

Singh, B. and Majumdar, A.J. (1981) Properties of GRC containing inorganic fillers, *International Journal of Cement Composites and Lightweight Concrete*, Vol. 3, pp. 93–102.

Singh, B. and Majumdar, A.J. (1985) The effect of PFA addition on the properties of GRC, *International Journal of Cement Composites and Lightweight Concrete*, Vol. 7, pp. 3–10.

Singh, B. and Majumdar, A.J. (1987) GRC made from supersulphated cement: 10 year results, *Composites,* Vol. 18, pp. 329–33.

Singh, B., Majumdar, A.J. and Ali, M.A. (1984) Properties of GRC containing PFA, *International Journal of Cement Composites and Lightweight Concrete*, Vol. 5, pp. 65–74.

Stucke, M.J. and Majumdar, A.J. (1976) Microstructure of glass fibre reinforced cement composites, *Journal of Materials Science*, Vol. 11, pp. 1019–30.

Studinka, J. (1986) Replacement of asbestos in the fiber cement industry - state of substitution, experience up to now. *International Man Made Fibres Congress*, Dornbirm, Austrian Chemical Institute, Paper 25.

Tanaka, M. and Uchida, I. (1985) Durability of GFRC with calcium silicate C_4A_3S-CS-slag type low alkaline cement, *Proceedings of Durability of Glass Fiber Reinforced Concrete Symposium,* (ed. S. Diamond), Prestressed Concrete Institute, Chicago, pp. 305–14.

Wang Jia *et al.*, (1986) Improvement of paste–aggregate interface by adding silica fume, *Proceedings of the 8th International Congress on the Chemistry of Cement*, Rio de Janeiro, Vol. 3. pp. 460–5.

West, J.M., De Vekey, R.C. and Majumdar, A.J. (1985) Acrylic-polymer modified GRC, *Composites, Vol.* 16, pp. 33–8.

West, J.M., Majumdar, A.J. and De Vekey, R.C. (1986) Properties of GRC modified by vinyl emulsion polymers, *Composites,* Vol. 17, pp. 56–62.

Wu Xuequan, Han Sufen, Brian Quichan and Tang Mingshu (1986) Effect of Pretreatment of aggregate surface on the properties of concrete, *Proceedings of the 8th International Congress on the Chemistry of Cement*, Rio de Janeiro, Vol. III, pp. 454–9.

Zimbelmann, R. (1987) A method for strengthening the bond between cement stone and aggregates, *Cement and Concrete Research*, Vol. 17, pp. 651–60.

PROPERTIES OF TRANSITION ZONE

3

Tests to determine the mechanical properties of the interfacial zone

S. Mindess

3.1 Introduction

Modern concretes contain a number of different interfaces, including those between

(i)   the various phases in the hydrated cement paste (hcp) system;
(ii)   the hcp and anhydrous cement grains;
(iii) the hcp and pozzolanic additives or finely divided mineral fillers;
(iv)   the hcp and aggregate particles;
(v)   the hcp or mortar and fibres in fibre-reinforced concrete (frc); and
(vi)   the concrete and steel reinforcing bars or prestressing cables.

While all of these are important, in this review only two types of interfaces are considered: those between hcp and aggregates, and those between hcp and fibres. The specific focus is on the complexities of the test methods that might be used to try to measure the mechanical properties of these interfaces. In large part because of these complexities, there are neither test methods in any national testing standards that deal with the determination of the mechanical properties of interfaces, nor methods that deal with the determination of the thickness or extent of the interfacial zone.

It has long been known that the various interfaces in concrete can all have a profound effect on the properties of concrete. As Le Chatelier (1887) pointed out with respect to hcp over 100 years ago, "The final hardness will depend upon the internal cohesion of the crystals, and upon their natural adherence". That is, the strength of hcp depends upon both the forces within individual particles of hydration products ('internal cohesion'), and the interparticle forces ('adherence') which bond the hydration products to each other and to the anhydrous cement. Similarly, Sabin (1905) pointed out that "since cement mortars are usually employed to bind other materials together, it follows that the adhesive strength is of the greatest importance". Unfortunately, however, it has so far not been possible to quantify, except in a few special cases, the effect of the interfacial properties on concrete properties, or even to determine the bond strengths between the various components of concrete. As Sabin (1905) also noted, "on account of the difficulty of making tests of adhesive strength, however, the data concerning it are very meagre", a statement which is still largely true today.

Interfacial Transition Zone in Concrete. Edited by J.C. Maso. RILEM Report 11.
Published in 1996 by E & FN Spon, 2-6 Boundary Row, London SE1 8HN. ISBN 0 419 20010 X.

3.1.1 THE NATURE OF THE INTERFACE

There have been a number of recent reviews of the nature of the interfacial regions in concrete (Maso, 1980; Struble *et al.*, 1980; Bartos, 1982; RILEM Colloquium, 1982; Diamond, 1986; Massazza and Costa, 1986; Mehta, 1986; Scrivener and Pratt, 1986; Mindess and Shah, 1988), and a few attempts to relate the bond strength to the properties of concrete (Mindess *et al.*, 1986; Mindess, 1989). The characterization of the cement–aggregate interface is described in detail elsewhere in chapter 1 of this volume (Pratt and Scrivener, 1995). Over the years, a number of models of the interfacial zone have been suggested, such as those shown in Fig. 3.1 (Mehta, 1986; Langton and Roy, 1980; Bentur *et al.*, 1986). For the purposes of this chapter, the precise details of the interfacial zone are not important. It is sufficient merely to note that the thickness of the interfacial zone is generally taken to be about 50 μm, with the major differences from the bulk hcp occurring within the first 20 μm from the physical interface. Moreover, very often the weakest part of

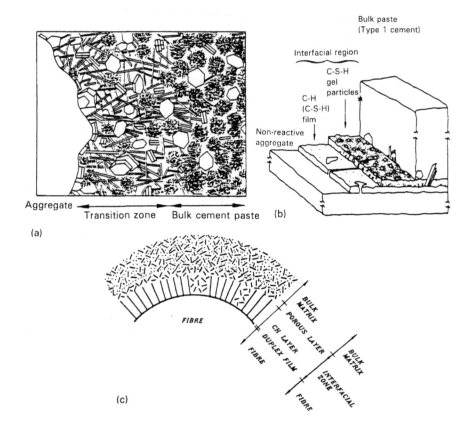

Fig. 3.1 (a) Representation of the transition zone in concrete (after Mehta, 1986; reproduced with permission); (b) schematic of the interfacial region between a non-reactive silica substrate and Type I cement paste (after Langton and Roy (1980); reproduced with permission); (c) schematic of the interfacial zone at the steel fibre–cement interface (after Bentur *et al.*, 1986).

the interfacial zone lies not right at the physical interface, but rather 5 to 10 μm within the paste fraction, with the fracture path running along the cleavage planes of the oriented $Ca(OH)_2$ crystals.

It is also important to recognize the extent of the 50 μm thick interfacial zone. It has been shown by Diamond *et al.* (1982) that, in concrete, the average spacing between adjacent aggregate particles is about 75–100 μm. Even though the variability in spacing is large, this suggest that a relatively large proportion of the hcp lies within the interfacial zone.

3.1.2 RELATIONSHIP BETWEEN THE PROPERTIES OF THE INTERFACE AND THE MECHANICAL PROPERTIES OF CONCRETE

Conventional wisdom has it that the cement–aggregate interfacial zone is the 'weak link' in concrete. Certainly, it is common to see the fracture paths in normal concrete lying largely along the interface between the hcp and the coarse aggregate particles. Whether this is due primarily to the inherent weakness of the interfacial zone, or perhaps also to stress concentrations induced by the more rigid aggregate particles embedded in the hcp matrix, is not entirely clear.

However, if we do assume that the cement–aggregate interface is the weakest part of concrete, then this gives rise to several questions:

1. What effects do the mechanical properties of the interface have on the mechanical properties of concrete?
2. What is the relative importance of the interfacial properties in determining the mechanical properties of concrete, when compared to the well-known effect of porosity?
3. Is it possible to alter the interfaces in a controlled way in order to improve the properties of concrete?

Most studies of strength show that increasing the cement–aggregate bond strength increases the concrete strength, whether in tension, compression, or flexure. However, these increases, while significant, are not overwhelming. In going from 'no bond' to 'perfect bond', strength increases generally lie in the range of 15 to 40%, with improvements in tensile strength being somewhat higher than those in compressive strength. A simple relationship between bond strength and either compressive or flexural strength was developed by Alexander and Taplin (1962, 1964), based on a regression analysis of the data then available. It has the form

$$\sigma = b_0 + b_1 m_1 + b_2 m_2$$

where σ = concrete strength (compression or flexure)
b_0, b_1, b_2 = linear regression coefficients
        = 480, 2.08, 1.02, respectively, for compression
        = 290, 0.318, 0.162, respectively, for flexure
m_1, m_2 = modulus of rupture of the paste and of the cement–aggregate bond, respectively.

From this expression, it may be seen that a change in the flexural strength of the paste has about twice as much effect as does a change in the flexural bond strength. Similar results were obtained by Darwin and Slate (1970), who found that large reductions in the cement–aggregate bond strength caused only small reductions in the modulus of elasticity and compressive strength of concrete. They concluded that "the strength of the interfacial bond is apparently of much less importance in the behaviour of concrete than has been thought in the past".

It should be emphasized here that these and other similar studies have all dealt with normal strength concretes. Though there are no systematic data available, it would appear that the cement–aggregate bond strength is of much more importance for the current generation of high strength concretes. Indeed, Aitcin (1991) has suggested that high strength concrete should be considered as a three-phase material: hardened cement paste, aggregate, and the transition zone.

In general, as the strength of concrete increases, the fracture path increasingly begins to go *through*, rather than around, aggregate particles; several studies have shown an increase in the percentage of broken aggregate particles as concrete strength (or the strength of the cement–aggregate bond) increases (Conjeaud *et al.*, 1980; Cottin *et al.*, 1982). This has two principal consequences: (i) the fracture toughness of the concrete increases, since aggregate particles tend to be tougher than the hcp; and (ii) the brittleness of the concrete also increases, probably because extensive microcracking in the interfacial zone begins only at a much higher relative stress level.

Thus, while the above discussion makes clear the importance of the interface in determining concrete properties, interfacial effects should nonetheless not be overemphasized. For instance, from fracture mechanics considerations, the strength of real concrete is governed by flaws of the order of size of 0.1 to 1 mm and larger. Similarly, the transport properties of concrete depend not only upon diffusion through the hcp (or the interface) but also upon the ingress of water, gases and ionic species through macroscopic cracks induced by plastic shrinkage or drying shrinkage. Such flaws and cracks are much larger than the 50 μm thick interfacial zone. Thus, careful laboratory studies on small specimens which have been protected against shrinkage probably overestimate the influence of the transition zone on the properties of the composite material.

3.2 Problems of specimen preparation for the measurement of interfacial mechanical properties

In order to determine the properties of the cement–aggregate interfacial region, clearly a specimen of some kind is needed. However, while it is now easy enough, with modern techniques, to examine the interfacial zone using SEM or X-ray diffraction techniques, determining the mechanical properties of the interfacial zone poses special problems, for which there are no obvious solutions. The first basic problem is that it is not possible to isolate the material (primarily hcp) in the transition zone for mechanical testing. Rather, we must test a composite specimen, consisting of coarse aggregate, hardened cement paste (or mortar) and the ~50 μm thick transition zone. The hcp blends gradually and smoothly into the transition zone, providing a 'perfect' bond between the two. However, the contact between the transition zone and the aggregate is never 'perfect', and is almost certainly not continuous. Thus, under any form of loading, the state of stress within the transition zone can never be known exactly, since we cannot define the boundary conditions with any certainty.

Even when we accept the necessity of preparing composite specimens, the question of just how we should prepare these specimens arises. The second basic problem is that, as with most mechanical tests on concrete, the numerical values obtained depend upon such factors as specimen size, specimen geometry, method and rate of loading, the moisture content of the specimen, the test temperature, and so on, as has been discussed in detail by Mindess and Young (1981). Since these problem are common to all concrete testing, they will not be discussed further here. However, this implies that unless some standard test methods for studying interfacial properties can be developed, it will not be possible to compare the numerical values of mechanical properties measured in different laboratories by different methods. In what follows in this section, we will deal primarily with the difficulties inherent in specimen preparation, regardless of the specimen geometry.

3.2.1 ROUGHNESS OF THE AGGREGATE SURFACE

The bond between hcp and rock must depend upon some combination of chemical bonding, physical bonding due to van der Waals' forces, and mechanical interlock, with the relative importance of these three mechanisms varying with different rock types and cement compositions. Clearly, however, the amount of mechanical interlock will depend upon the roughness of the aggregate surface. Some investigators, such as Odler and Zürz (1988), try to eliminate possible interlocking effects by polishing the rock surface to a 'mirror-like' finish. Others (e.g. Struble and Mindess, 1983) have ground the surfaces with a particular size of grit, though this will create a different surface roughness with different rock types. Another technique adopted by Struble (1988) is to cast the cement against the surface obtained by cutting with a low-speed diamond saw. More recently, Alexander *et al.* (1992) have cast cement against the rather rough fracture surface of rock. There is no 'right' or 'wrong' method of surface preparation, but obviously these different surface finishes will provide quite different mechanical bond properties.

3.2.2 VARIABILITY WITHIN THE TRANSITION ZONE

As indicated earlier, the interfacial zone, though only about 50 μm thick, is very variable in its composition (Fig. 3.1) and thus its mechanical properties must vary also, as indicated by the microhardness data of Lyubimova and Pinus (1962) shown in Fig. 3.2. Thus, any strength measurements will depend on just where within the transition zone failure occurs, and this will in turn depend on the particular cement–aggregate system that we are testing. This point has been illustrated very clearly by Odler and Zürz (1988); Fig. 3.3 shows the different mechanisms of failure that they found for different systems. These differences in failure mechanism would make it very difficult to assign a particular strength (or elastic modulus) to the interfacial zone. We can at best determine only the strength of the weakest plane in the interfacial zone for a particular system, or an 'average' elastic modulus.

3.2.3 BLEEDING

Another problem in preparing specimens for interfacial studies is that of bleeding. Bleeding arises because, under the force of gravity, the solid aggregate particles in suspension in fresh concrete tend to settle, since they have a higher density than the fluid paste. In practice, bleed water is trapped at the undersides of the larger aggregate particles (Fig. 3.4, Mehta,

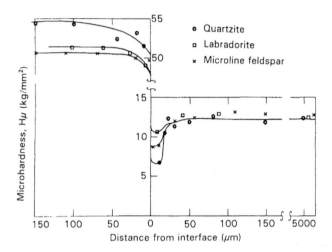

Fig. 3.2 Variation in microhardness across the cement–aggregate interface. Based on data from Lyubimova and Pinus (1962).

1986). These lenses of bleed water are frequent, particularly in the upper part of any concrete lift. Thus, if we try to study interfaces *in situ*, they will vary, in the general case, from the bottom to the top of a coarse aggregate particle. Related to this, Zimbelmann (1987) described the way in which a water film could form around aggregate particles. In his studies, he could find no cement grains within about 15 to 25 µm of the rock surface. This

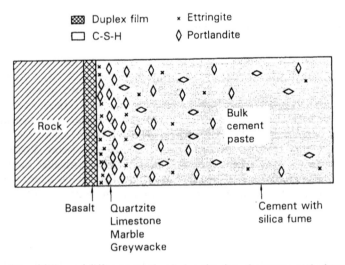

Fig. 3.3 Mechanism of failure of different samples during cleaving; the arrows at the bottom indicate the location of the failure plane for different rock types. After Odler and Zürz (1988); reproduced with permission.

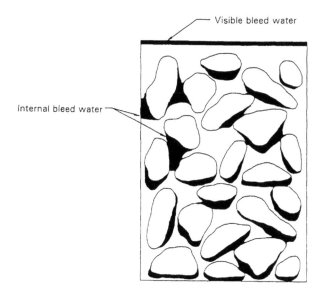

Fig. 3.4 Bleed water in concrete. After Mehta (1986); reproduced with permission.

would lead to the development of ~ 10 μm thick water films on the solid particle (Fig. 3.5). For specimens prepared for interfacial studies by casting cement paste against a large piece of rock, it is thus very difficult to avoid the development of a water-rich layer right at the rock face, though if the specimens are rotated in a vertical pane slowly but continuously until setting has taken place, this problem can at least be minimized.

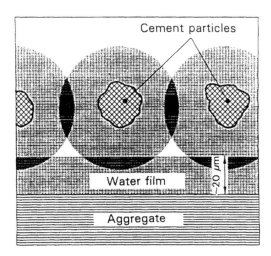

Fig. 3.5 Formation of a water film during the casting of cement mortar against coarse aggregate; after Zimbelmann (1987); reproduced with permission.

3.2.4 COMPOSITE SAMPLES VERSUS INTERFACES IN REAL CONCRETE

Finally, there appear to be some differences between the interfaces occurring in concrete mixed in a conventional way, and those which result from the casting of cement paste against a rock surface. For instance, Scrivener and Pratt (1986) have stated that "the relative movement of the sand and cement grains during mixing, and possibly settling of the sand grains before the cement paste sets, may lead to regions of low paste density around the sand grains and to areas of localized bleeding at the interface in which large CH crystals precipitate." Further, Scrivener *et al.* (1988) have argued "that the presence of the aggregate particles during mixing may affect the gradients of microstructure in [the interfacial] zone, but has little effect on its width".

In summary then, however carefully we try to prepare specimens for interfacial studies, we will always have to live with the fact that the results are essentially artifact of the details of specimen preparation and handling. It is extremely difficult, if not impossible, to simulate in the laboratory the interfaces that occur in commercially produced concrete.

3.3 Measurements of interfacial properties

3.3.1 MECHANISMS OF FAILURE

Before dealing with the measurement of bond strength, it is first necessary to describe briefly the mechanisms of failure. It has long been known that, because the aggregate particles have quite different elastic properties from those of the paste, and because the cement–aggregate bond is imperfect, when concrete is loaded highly localized stresses and strains occur in the transition zone in the vicinity of the larger aggregate particles. Dantu (1958), using photoelastic techniques, found that the localized strains were about 4.5 times the average strains, and localized stresses were about twice as high as the average stresses. This helps to account for the fact that initial cracking occurs at the interface.

Vile (1968) developed a model for the stress distribution around a single aggregate particle in a cement matrix, under compressive stress. For the usual case of the aggregate particle being stiffer than the paste, the apparent order of failure is (1) tensile bond failure; (2) shear bond failure; (3) shear and tensile matrix failure; and (4) occasional aggregate failure.

3.3.2 BOND STRENGTH

The most common tests used to try to determine the mechanical properties of the interfacial zone are those for the aggregate–cement bond strength. A wide variety of test configurations has been used; some of the more common ones are shown in Fig. 3.6, taken from Alexander *et al.* (1968). All the experimental test arrangements shown are, unfortunately, prone to the problems described earlier with regard to preparing cement–aggregate composites.

Early studies of bond strength revealed that the bond strength was related to the compressive strength of the paste, though there is enough variability with different brands of nominally the same type of cement to suggest that bond strength is also affected by the details of the chemical composition of the cement. This chemical interaction, which is not yet understood, further complicates studies of bond strength. Over time, the bond strength becomes essentially equal to the paste strength. In addition, the bond strength is related to

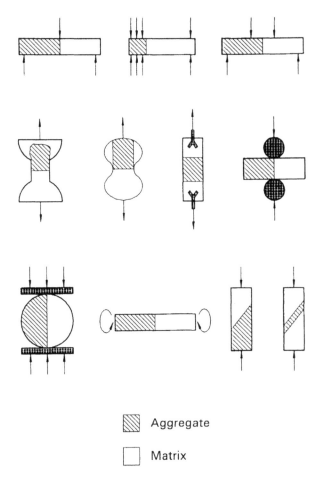

Fig. 3.6 Techniques used for measuring aggregate–cement bond strength. After Alexander *et al.* (1968).

the type of rock, with siliceous rocks developing the highest strengths. For extrusive rocks, the bond strength appeared to increase with an increase in silica content. Overall, Alexander *et al.* (1968) reported that the 'best' rocks might develop about twice as much bond strength as the 'worst' rocks; Odler and Zürz (1988) obtained similar results.

Zimbelmann (1985) concluded, from studies of the tensile strength of the interface conducted using the specimen shown in Fig. 3.7, that the adhesion between the cement and the aggregate depends mainly on the effect of physical, rather than chemical forces. Limestone appeared to be distinctly different from other aggregates, presumably because of the epitaxial growth on the limestone surface. Zimbelmann (1987) later developed a method of increasing the bond strength, either by the use of a detergent to reduce the porosity at the interface, or by the use of a suspension of pozzolanic material in water glass. This led to approximately a three-fold increase in bond strength.

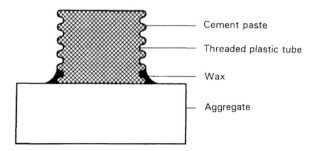

Fig. 3.7 Sample for testing bond strength between hcp and aggregate; after Zimbelmann (1985); reproduced with permission.

3.3.3 INDIRECT MEASURES OF BOND STRENGTH

In order to try to avoid the difficulties inherent in casting cement pastes against rock surfaces, a number of investigators have tried to deduce the effects of bond strength on concrete properties from tests on concrete specimens prepared with aggregates artificially treated in order to alter their bond characteristics. Here too the results have been rather contradictory. Hsu *et al.* (1963) and Perry and Gillott (1977) found little effect of bond strength on compressive strength. However, Hsu *et al.* (1963) did suggest that increasing the bond strength would increase the stress level at which extensive microcracking began. This was confirmed by Patten (1972), who indicated that a poor bond allows cracks to propagate more rapidly, hastening failure. On the other hand, more recent work by Chen and Wang (1988) showed that an improvement in interfacial bond strength would increase the tensile strength, and to a lesser extent the compressive strength, of concrete. Similar results were obtained by Wu *et al.* (1988), and by Wu and Zhou (1988). In work referred to earlier, Maso (1980) also found that the bond strength had a larger effect on tensile strength than on compressive strength.

It is difficult to reconcile these results. While some studies show that increases in bond strength increase tensile, flexural and compressive strengths, others do not, and indeed Fagerlund (1973) has argued, upon theoretical grounds, that the cement–aggregate bond should have little effect on compressive strength. It can only be said that these contradictory results confirm the difficulty of carrying out test on bond strength.

3.3.4 MICROHARDNESS

One of the few test methods which is able to measure directly at least some mechanical property of the interfacial zone is the microhardness test. In this test, the size of an indentation is measured as a function of the applied load, and an average stress over the indented area can be calculated. These tests have been used by a number of investigators over the years (Lyubimova and Pinus, 1962; Wang, 1988; Mehta and Monteiro, 1988; Wei *et al.*, 1986; Saito and Kawamura, 1986). As with other indirect measures of strength, the values obtained depend upon the type of equipment used (i.e. the geometry of the indenter), and upon the surface preparation. In order to scan across the interfacial zone with microhardness tests, very small loads are used, in order to create very small indentation, typically with dimensions of a few microns. These tests are, therefore, very sensitive to the smoothness of the surface, which in turn depends upon the polishing technique. As a consequence, 'hardness' values obtained by various investigators for various systems cover a wide range:

Mehta and Monteiro (1988)	~ 1 kg/mm^2
Lyubimova and Pinus (1962)	10–15 kg/mm^2
Wang (1988)	10–30 kg/mm^2
Wei *et al.* (1986) (cement–steel interface)	20–40 kg/mm^2

However, all of the investigators are in agreement that the weakest (softest) part of the interfacial zone lies about 20 μm into the paste, measuring from the aggregate (or steel) surface.

Microhardness tests do involve only the interfacial zone if properly carried out. Unfortunately, for the very heterogeneous structure of the interface on a microscale, it is not possible to convert microhardness measurements into any sort of meaningful strength values.

3.3.5 FRACTURE MECHANICS PROPERTIES

A number of investigators have tried to determine the fracture mechanics parameters of the interfacial zone. These tests have mostly involved the preparation of fracture specimens by casting cement paste against rock, and then testing the resulting composite specimens, usually in flexure. For instance, Hornain *et al.* (1982) used the specimen geometry shown in Fig. 3.8, and found that at the age of 28 days, the fracture toughness of the interface was about the same as that of the hardened cement paste. A similar specimen was used by Wang *et al.* (1986), who found that densifying the interface with the addition of silica fume increased the fracture energy of the interface considerably. More recently, Alexander (1991) has applied the ISRM Method I (1988) test of chevron-notched specimens to study interfacial fracture properties, and further work of this type has been carried out by Alexander *et al.* (1992). It was found by Alexander (1991) that interfacial fracture energy for paste–andesite specimens was higher than that of the paste alone, while the fracture energy at the cement paste–dolomite interface was lower than that of the paste.

Clearly, the problems described in detail earlier with preparing cement paste–rock specimens apply also to tests of fracture parameters, and so it is difficult to draw any general conclusions from the fracture studies mentioned above. A detailed review of the difficulties in preparing and testing such fracture specimens has been presented by Ziegeldorf (1983).

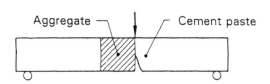

Fig. 3.8 Specimen used to measure the fracture toughness of the cement–aggregate interface; after Hornain *et al.* (1982) and Wang *et al.* (1986).

3.3.6 BOND STRENGTH BETWEEN HARDENED CEMENT PASTE AND FIBRES

A great many researchers have studied the fibre–cement interface, and the problem of the bond strength between fibres and cements. Typically, microhardness tests (e.g. Wei *et al.*, 1986), and fibre pull-out tests have been carried out.

A detailed review of the types of the types of pull-out tests that have been used has been presented by Gray (1983). Gray classified the tests into four categories, as shown schematically in Fig. 3.9:

(i)   single-fibre, single-sided,
(ii) single-fibre, double-sided,
(iii) multiple-fibre, single-sided, and
(iv) multiple-fibre, double-sided.

The details of some of the various test that have been attempted are shown in Fig. 3.10.

In mechanical tests of fibre–cement composites, two sets of problems arise. The first set is similar to those encountered with cement–rock composites, involving bleeding effects around the fibres, and the fact that embedding fibres carefully in a cement matrix does not simulate the sort of interfacial structure that occurs during conventional mixing of fibre-reinforced cement or concrete.

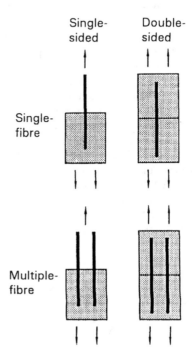

Fig. 3.9 Types and configurations of fibre-matrix bond test specimens; after Gray (1983); reproduced with permission.

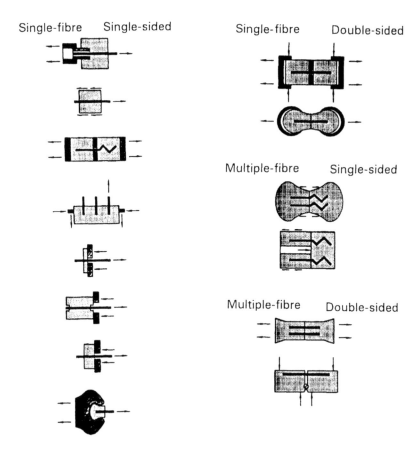

Fig. 3.10 Experimental techniques used to measure the fibre–matrix bond. After Gray (1983); reproduced with permission.

The second set of problems arises from the stress and strain conditions in the test specimens. Ideally, the stress and strain conditions at the fibre–matrix interface should be the same as those in the fractured composite material during fibre debonding and fibre withdrawal at a crack surface. Gray (1983) concluded that none of the methods shown in Fig. 3.10 is ideal, though clearly some were better than others. He suggested that two standard specimens be developed, a single-fibre specimen (because of its simplicity) for general use, and a multiple-fibre specimen that would be used occasionally when more precise value are required. A further problem is that most modern steel fibres are deformed in one way or another, and are not symmetrical about their longitudinal axis. It is therefore almost impossible to devise a test method for these deformed steel fibres for which the stress conditions at the interface can be easily defined. The problem is even more intractable for test with fibrillated polypropylene fibres, or multi-strand glass fibres.

3.4 Elastic properties

Because of the small thicknesses and lack of uniformity of the interfacial zone, there is no direct method for measuring the elastic properties of this zone. It may be possible to deduce, for instance, the modulus of elasticity of the interface from beam tests of composite specimens, for which the elastic moduli of the paste and rock are known separately. However, this involves a number of approximations (size of the interfacial zone, how perfect the cement–rock interface is, etc.), and is thus not likely to yield useful results.

Recently, Alexander (1992) has adopted a novel approach. He has prepared composite compression cylinders, consisting of a 'sandwich' structure of from 7 to 11 alternating layers of rock and a cement or mortar matrix, the thickness of each layer being about 15 mm. By measuring the elastic modulus in compression of these specimens, as well as those of the rock and matrix materials separately, the influence of the interfacial zone can be estimated. His data for elastic modulus vs age for such specimens agree reasonably well with theoretical calculations of the elastic modulus based on a simple series model.

3.5 Suggestions for further research

From the above, it is clear that there are many unsolved problems with regard to the measurement of the mechanical properties of the interfacial zone. There are a number of research directions which might be pursued:

1. Most important would be agreement on a standard test specimen (or specimens) and a standard method of specimen preparation and testing;
2. Methods of minimizing bleeding at the interface;
3. Analytical methods for predicting the mechanical properties of concrete from a knowledge of the properties of the cement, rock (or fibre) and the interfacial zone.

3.6 Conclusions

After decades of studying the interfacial zone, the microstructural characteristics of the various interfacial regions are reasonably well known. However, there is still no agreement as to how to determine the mechanical properties of the interfacial zone. The many measurements of these properties that have been carried out cannot readily be compared with each other, because of differences in test procedures. We are thus still very far from the goal of being able to use the properties of the interfacial region to help predict the properties of concrete.

Acknowledgements

This work was supported by the Natural Sciences and Engineering Research Council of Canada, though the Network of Centres of Excellence on High-Performance Concrete.

3.7 References

Aitcin, P.-C. (1991) Private communication.

Alexander, K.M. and Taplin, J.H. (1962) Concrete strength, paste strength, cement hydration and maturity rule, *Australian Journal of Applied Science,* Vol. 13, No. 4, pp. 277–84.

Alexander, K.M. and Taplin, J.H. (1964) Analysis of the strength and fracture of concrete based on unusual insensitivity of cement–aggregate bond to curing temperature, *Australian Journal of Applied Science,* Vol. 15, No. 3, pp. 160–70.

Alexander, K.M., Wardlaw, J. and Gilbert, D.J. (1965) Aggregate–cement bond, cement paste strength and the strength of concrete, *The Structure of Concrete and its Behaviour under Load,* (eds Brooks, A.E. and Newman, K.), Cement and Concrete Association, London, pp. 59–61.

Alexander, M.G. (1991) Fracture energies of interfaces between cement paste and rock, and applications to the engineering behaviour of concrete, *Fracture Processes in Brittle Disordered Materials,* (eds J.G.M van Mier *et al.*), RILEM International Symposium, Noordwijk, Netherlands, E & FN Spon, London, pp. 337–46.

Alexander, M.G. (1993) Two experimental techniques for studying the effects of the interfacial zone between cement paste and rock, *Cement and Concrete Research,* Vol. 13, No. 3, May, pp. 567–75.

Alexander, M.G., Mindess, S. and Qu, L. (1992) The influence of rock and cement types on the fracture properties of the interfacial zone, *Interfaces in Cementitious Composites,* (ed. J.C. Maso), RILEM International Conference, Toulouse, France, E & FN Spon, London, pp. 129–37.

Bartos, P.J.M. (ed.) (1982) *Bond in Concrete,* Applied Science Publishers, London.

Bentur, A., Gray, R.J. and Mindess, S. (1986) Cracking and pull-out process in fibre reinforced cementitious materials, (eds R.N. Swamy, R.L. Wagstaffe and D.R. Oakley), *Developments in Fibre Reinforced Cement and Concrete,* RILEM, Vol. 2, Paper 6.1.

Chen, Zhi Yuan and Wang, Jian Guo (1988) Effect of bond strength between aggregate and cement paste on the mechanical behaviour of concrete, *Bonding in Cementitious Composites,* (eds Mindess, S. and Shah, S.P.), Materials Research Society, Vol. 114, pp. 41–7.

Conjeaud, M., Lelong, B. and Cariou, B. (1980), Liaison pate de ciment Portland - granulets naturel, *Proceedings of 7th International Congress on the Chemistry of Cement,* Vol. III, Editions Septima, Paris, pp. VII-6 to VII-11.

Cottin, B., Marcdargent, S. and Cariou, B. (1982) Reactions between active aggregates and hydrated cement paste, *International RILEM Colloquium, Liaisons Pâtes de Ciment/Matériaux Associés,* Toulouse, France, pp. C20–C26.

Dantu, P. (1958) Etude des contraintes dans les mileux heterogenes. Application au beton, *Annales de L'Institut Technique due Batiment et des Travaux Publiques,* Vol. 11, No. 121, pp. 55–77.

Darwin, D. and Slate, F.O. (1970) Effect of paste–aggregate bond strength on behaviour of concrete, *ASTM Journal of Materials,* Vol. 5, No. 1, pp. 86–98.

Diamond, S. (1986) The microstructure of cement paste in concrete, *Proceedings of 8th International Congress on the Chemistry of Cement,* Vol. I, Finep, Rio de Janeiro, Brazil, pp. 122–47.

Diamond, S., Mindess, S. and Lovell, J. (1982) On the spacing between aggregate grains and the dimension of the aureole de transition, *International RILEM Colloquium, Liaisons Pâtes de Ciment/Matériaux Associés,* Toulouse, France, pp. C42–C46.

Fagerlund, G. (1973) Strength and porosity of concrete, *Proceedings of the International Symposium on Pore Structure and Properties of Materials,* RILEM/IUPAC, Prague, pp. D51–D73.

Gray, R.J. (1983) Experimental techniques for measuring fibre/matrix interfacial bond shear strength, *Testing, Evaluation and Quality Control of Composites,* (ed. Feest, T.), Butterworth Scientific Ltd, UK, pp. 3–11.

Hornain, H., Mortureux, B. and Regourd, M. (1982) Aspects physico-chimique et mecanique de la liaison pate de ciment-granulat, *International RILEM Colloquium, Liaisons Pâtes de Ciment/Matériaux Associés,* Toulouse, France, pp. C56–C65.

Hsu, T.T.C., Slate, F.O., Sturman, G.M. and Winter, G. (1963) Microcracking of plain concrete and the shape of the stress–strain curve, *Journal of the American Concrete Institute, Proceedings,* Vol. 60, No. 2, pp. 209–24.

International Society for Rock Mechanics (1988) Suggested methods for determining the fracture toughness of rock, *International Journal of Rock Mechanics and Mineral Science*, Vol. 25, No. 2, pp. 71–96.

Langton, C.A. and Roy, D.M. (1980) Morphology and microstructure of cement paste/rock interfacial regions, *Proceedings of 7th International Congress on the Chemistry of Cement*, Vol. III, Editions Septima, Paris. pp. VI-127 – VII-132.

Le Chatelier, H. (1887) *Experimental Researches on the Constitution of Hydraulic Mortars*, translated by J.L. Mack, McGraw Publishing Company, New York, 1905.

Lyubimova, T.Yu and Pinus, E.R. (1962) Crystallization structure in the contact zone between aggregate and cement in concrete, *Kolloidnyi Zhurnal* (USSR), Vol. 24, No. 5, pp. 578–87.

Maso, J.C. (1980) The bond between aggregates and hydrated cement paste, *Proceedings of 7th International Congress on the Chemistry of Cement*, Vol. I, Editions Septima, Paris, pp. VII-1/3 – VII-1/15.

Massazza, F. and Costa, U. (1986) Bond: paste–aggregate, paste–reinforcement and paste–fibres, *Proceedings of 8th International Congress on the Chemistry of Cement*, Vol. I, Finep, Rio de Janeiro, Brazil, pp. 158–80.

Mehta, P.K. (1986a) *Concrete: Structure, Properties and Materials*, Prentice-Hall, Englewood Cliffs, NJ.

Mehta, P.K. (1986) Hardened cement paste - microstructure and its relationship to properties, *Proceedings of 8th International Congress on the Chemistry of Cement*, Finep, Rio de Janeiro, Brazil, Vol. I, pp. 113–21.

Mehta, P.K. and Monteiro, P.J.M. (1988) Effect of the aggregate cement, and mineral admixtures on the microstructure of the transition zone, (eds Mindess, S. and Shah, S.P.), *Bonding in Cementitious Composites*, Materials Research Society, Vol. 114, pp. 65–75.

Mindess, S. (1989) Interfaces in concrete, (ed. J. Skalny), *Materials Science of Concrete*, American Ceramic Society, pp. 163–80.

Mindess, S., Odler, I. and Skalny, J. (1986) Significance to concrete performance of interfaces and bond: challenges of the future, *Proceedings of 8th International Congress on the Chemistry of Cement*, Finep, Rio de Janeiro, Brazil, Vol. I, pp. 151–157.

Mindess, S. and Shah, S.P. (eds) (1988) *Bonding in Cementitious Composites*, (eds S. Mindess and S.P. Shah), Materials Research Society, Vol. 114.

Mindess, S. and Young, J.F. (1981) *Concrete*, Prentice-Hall, Englewood Cliffs, NJ, pp. 410–37.

Odler, I. and Zürz, A. (1988) Structure and bond strength of cement-aggregate interfaces, (eds Mindess, S. and Shah, S.P.), *Bonding in Cementitious Composites*, Materials Research Society, Vol. 114, pp. 21–7.

Patten, B.J.F. (1972) The effects of adhesive bond between coarse aggregate and mortar on the physical properties of concrete, UNICIV Report No. R-82, University of New South Wales, Australia.

Perry, C. and Gillott, J.E. (1977) The influence of mortar–aggregate bond strength on the behaviour of concrete in uniaxial compression, *Cement and Concrete Research*, Vol. 7, No. 5, pp. 553–64.

Pratt, P.L. and Scrivener, K.L. (1995) Characterization of interfacial microstructure, *This volume.*

RILEM (1982) *International Colloquium, Liaisons Pâtes de Ciment/Matériaux Associés*, Proceedings, RILEM Colloquium, Toulouse, France.

Sabin, L.C. (1905) *Cement and Concrete*, Archibald Constable and Co., Ltd, London, pp. 272–3.

Saito, M. and Kawamura, M. (1986) Resistance of the cement–aggregate interfacial zone to the propagation of cracks, *Cement and Concrete Research*, Vol. 16, No. 5, pp. 653–61.

Scrivener, K.L., Crumbie, A.K. and Pratt, P.L. (1988) A study of the interfacial region between cement paste and aggregate in concrete, *Bonding in Cementitious Composites*, (eds Mindess, S. and Shah, S.P.), Materials Research Society Symposia, Vol. 114, pp. 87–9.

Scrivener, K.L. and Pratt, P.L. (1986) A preliminary study of the microstructure of the cement/sand bond in mortars, *Proceedings of 8th International Congress on the Chemistry of Cement*, Finep, Rio de Janeiro, Brazil, Vol. III, pp. 466–71.

Struble, L. (1988) Microstructure and fracture at the cement paste–aggregate interface, *Bonding in Cementitious Composites*, (eds Mindess, S. and Shah, S.P.), Materials Research Society, Vol. 114, pp. 11–20.

Struble, L. and Mindess, S. (1983) Morphology of the cement–aggregate bond, *International Journal of*

Cement Composites and Lightweight Aggregate, Vol. 5, pp. 79–86.

Struble, L., Skalny, J. and Mindess, S. (1980) A review of the cement–aggregate bond, *Cement and Concrete Research*, Vol. 10, No. 2, pp. 277–86.

Vile, G.W.D. (1968) The strength of concrete under short-term static biaxial stress, *The Structure of Concrete and its Behaviour under Load*, (eds Brooks, A.E. and Newman, K.), Cement and Concrete Association, London, pp. 275–88.

Wang, J., Liu, B., Xie, S. and Wu, Z. (1986) Study of the interface strength on steel fibre reinforced cement-based composites, *Journal of the American Concrete Institute, Proceedings*, Vol. 83, pp. 597–605.

Wang, Yuji (1988) The effects of bond characteristics between steel slag fine aggregate and cement paste on mechanical properties of concrete and mortar, *Bonding in Cementitious Composites*, (eds Mindess, S. and Shah, S.P.), Materials Research Society, Vol. 114, pp. 49–51.

Wei, S., Mandel, J.A. and Sard, S. (1986) Study of the interface strength on steel fibre reinforced cement-based composites, *Journal of the American Concrete Institute, Proceedings*, Vol. 83, pp. 597–605.

Wu, X., Li, D., Wu, X., and Tang, M. (1988) Modification of the interfacial zone between aggregate and cement paste, *Bonding in Cementitious Composites*, (eds Mindess, S. and Shah, S.P.), Materials Research Society, Vol. 114, pp. 35–40.

Wu, K. and Zhou, J. (1988) The influence of the matrix–aggregate bond on the strength and brittleness of concrete, *Bonding in Cementitious Composites*, (eds Mindess, S. and Shah, S.P.), Materials Research Society, Vol. 114, pp. 29–34.

Ziegeldorf, S. (1983) Fracture mechanics parameters of hardened cement paste, aggregates and interfaces, *Fracture Mechanics of Concrete*, (ed. Wittmann, F.H.), Elsevier, Amsterdam, pp. 371–409.

Zimbelmann, R. (1985) A contribution to the problem of cement-aggregate bond, *Cement and Concrete Research*, Vol. 15, No. 5, pp. 801–8.

Zimbelmann, R. (1987) A method for strengthening the bond between cement stone and aggregates, *Cement and Concrete Research*, Vol. 17, No. 4, pp. 651–60.

4

Mechanical modelling of the transition zone

P.J.M. Monteiro

4.1 Introduction

Traditionally, the composite nature of plain concrete has been modelled as a two-phase material (rigid inclusions of rocks dispersed in the cement paste matrix). This assumption is oversimplified, giving limited and sometimes questionable overall information. As described in detail in various chapters of this book, the transition zone existing between the aggregate and the cement paste is characterized by: (a) higher porosity than the matrix, (b) existence of larger crystals than in the matrix and (c) precipitation of calcium hydroxide with a preferred orientation. The realization of the importance of the transition zone was fundamental for the development of more durable and stronger concretes. Unfortunately however little has been done to incorporate the transition zone as a separate phase in the mathematical modelling of concrete as a composite material.

The lack of research makes this chapter different from the others, where a fairly comprehensive summary of the state-of-the-art is presented and by-and-large represents the consensus among researchers in the field. The material presented here reflects the author's point of view in a field not yet mature and which, hopefully, will change in the near future due to the development of truly realistic models.

The effect of the transition zone on the composite behaviour of concrete will be discussed in section 4.2, where the Hashin-Strikman (H-S) bounds will be used to assess whether concrete behaves as a two- or three-phase material in the elastic domain. Section 2 describes how the higher porosity at the transition zone can be modelled by a generalized continuum theory of the volume-fraction type.

4.2 Composite nature of concrete

The importance of microcracking at high stress levels was made clear by the pioneering work of Hsu *et al.* (1963). The results indicated that at stresses greater than 50% of the ultimate compressive load a complex system of microcracks would develop at the transition zone between the aggregate and the cement paste. Therefore, the importance of the transition zone on cracking mechanism and ultimate strength was obvious. Less noticeable was the relative importance of the transition zone in the elastic domain, e.g. for compressive stresses less than 40% of ultimate load. Indeed, it was taken for granted that its relevance was minimal so two-phase models could be used for concrete.

Figs 4.1 (a) and (b) show two of the most elementary models for two-phase materials. A significant development was the work of Hirsch (1962) and Dougill (1962), who proposed

Interfacial Transition Zone in Concrete. Edited by J.C. Maso. RILEM Report 11.
Published in 1995 by E & FN Spon, 2–6 Boundary Row, London SE1 8HN. ISBN 0 419 20010 X.

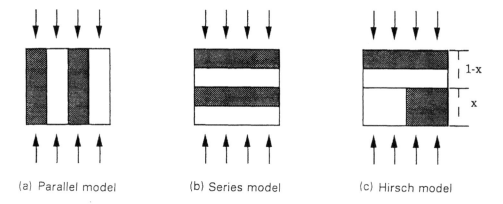

(a) Parallel model   (b) Series model   (c) Hirsch model

Fig. 4.1 Simple two-phase models for concrete.

a combined series-parallel model shown in Fig. 4.1 (c). The elastic modulus would be expressed by:

$$\frac{1}{E_c} = (1 - x)\left[\frac{c}{E_a} + \frac{1 - c}{E_m}\right] + x\left[\frac{1}{cE_a + (1 - c)E_m}\right] \qquad (4.1)$$

where

$$c = \frac{V_a}{V_c}$$

In this case x and $(1 - x)$ indicate the relative contributions of the parallel (uniform strain) and series (uniform stress) models. Often researchers have reported the use of the Hirsch model to estimate the degree of bonding between the cement paste matrix and aggregate in concrete. Monteiro (1991) pointed out that the uniform stress model does not imply *no bond* between the phases, and the uniform strain model does not indicate a *perfect bond*.

More elaborate two-phase models were developed over time. However, little research was undertaken to assess the theoretical applicability of a two-phase model for the description of concrete. To address this issue let us consider the work of Hashin-Strikman (H-S) (1963) who developed stringent bounds for a composite material which is, in a statistical sense, isotropic and homogeneous. The H-S bounds were derived using variational principles of the linear theory of elasticity for multiphase materials of arbitrary phase geometry. For two-phase composites the expressions take the form:

$$K_1^* = K_1 + \cfrac{v_2}{\cfrac{1}{K_2 - K_1} + \cfrac{3v_1}{3K_1 + 4G_1}}$$

$$K_2^* = K_2 + \cfrac{v_1}{\cfrac{1}{K_1 - K_2} + \cfrac{3v_2}{3K_2 + 4G_2}}$$

$$G_1^* = G_1 + \cfrac{v_2}{\cfrac{1}{G_2 - G_1} + \cfrac{6(K_1 + 2G_1)v_1}{5G_1(3K_1 + 4G_1)}}$$

$$G_2^* = G_2 + \cfrac{v_1}{\cfrac{1}{G_1 - G_2} + \cfrac{6(K_2 + 2G_2)v_2}{5G_2(3K_2 + 4G_2)}}$$

where K and G are the bulk and shear moduli respectively. The volume fractions of phases 1 and 2 are expressed by v_1 and v_2. Here $K_2 > K_1$; $G_2 > G_1$.

If a composite material behaves as a two-phase continuum, it should satisfy the H-S bounds. The author collected some original data from a previous research study (Monteiro, 1986) on elastic moduli of mortar as affected by the volume of sand concentration. Fig. 4.2 shows that the experimental results for sand concentrations above 0.3 lay outside the bounds indicating that a two-phase model is not adequate for mortar.

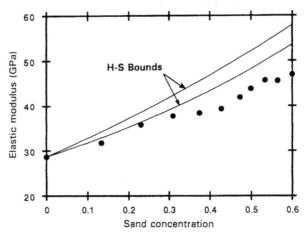

Fig. 4.2 Elastic modulus of mortar as a function of sand concentration.

Table 4.1 shows the results reported by Nilsen and Monteiro (1993) on measured elastic modulus of concrete and how they compare to the upper and lower H-S bounds. From Table 4.1 it is seen that when lead, limestone and to some extent gravel, are used as aggregate, the measured elastic modulus values are lower than the lower H-S bound. Fig. 4.3 shows an

Table 4.1. Volumetric mix proportions, results of elastic moduli measurements, and calculations of the Hashin-Strikman bounds, * after Hirsch (1962).

Batch designation*	Aggregate type*	Ratio of absolute volume of aggregate to volume of concrete* V_a/V_c	Elastic moduli of paste* (GPa)	Elastic moduli of aggregate* (GPa)	Measured elastic moduli* (GPa)	Lower H-S bound (GPa)	Upper H-S bound (GPa)
CM	-		19.17				
ST-5	Steel	0.5		206	53.4	46.4	97.7
ST-4	Steel	0.4		206	44.5	39.2	80.1
ST-3	Steel	0.3		206	37.3	33.3	63.5
SD-4	Sand	0.4		75.8	33.9	31.2	37.4
SD-2	Sand	0.2		75.8	25.7	24.8	27.8
GL-6	Glass	0.57		73.1	42.8	37.9	45.1
GL-5	Glass	0.5		73.1	37.4	35.0	41.4
GL-4	Glass	0.4		73.1	35.3	31.1	36.4
GL-3	Glass	0.3		73.1	30.8	27.7	31.7
GL-2	Glass	0.2		73.1	26.1	24.6	27.3
GR-6	Gravel	0.57		59.6	36.1	35.5	40.3
GR-5	Gravel	0.5		59.6	32.5	33.0	37.3
GR-4	Gravel	0.4		59.6	29.7	29.7	33.2
GR-3	Gravel	0.3		59.6	27.4	26.7	29.4
GR-2	Gravel	0.2		59.6	23.0	24.0	25.9
LI-6	Limestone	0.57		30.7	23.4	25.5	25.9
LI-5	Limestone	0.5		30.7	23.9	24.7	25.0
LI-4	Limestone	0.4		30.7	22.3	23.4	23.8
LI-3	Limestone	0.3		30.7	20.6	22.3	22.6
LI-2	Limestone	0.2		30.7	20.2	21.2	21.4
PB-5	Lead	0.5		16.6	16.2	17.0	17.0
PB-4	Lead	0.4		16.6	16.7	17.4	17.4
PB-2	Lead	0.2		16.6	17.2	18.3	18.3

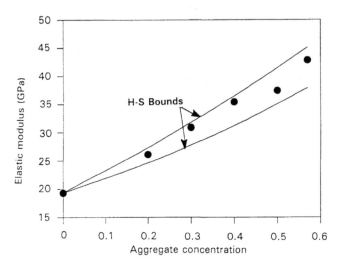

Fig. 4.3 Elastic modulus of mortar as a function of glass aggregate concentration.

example when the elastic modulus lies within the H-S bounds, while Fig. 4.4 shows an instance when it violates the H-S bounds. The fact that the experimental results are outside the bound indicates that modelling concrete as a two-phase material is not appropriate. Stiffer materials (i.e. sand, glass and steel punchings) when used as aggregates make the measured elastic modulus to be within the H-S bounds because the width of the H-S bounds increases with the ratio E_a/E_{cm}, where E_a and E_{cm} are the elastic modulus of the aggregate and cement paste respectively.

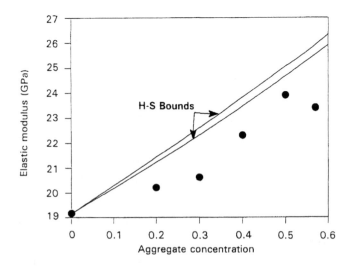

Fig. 4.4 Elastic modulus of mortar as a function of limestone aggregate concentration.

4.3 A generalized continuum theory for concrete

The porosity gradient existing in the transition zone can be incorporated in a generalized continuum if introduced as an additional variable that is not included in the traditional formulations of continuum mechanics. A model describing a deformable body is called a generalized continuum if, besides the displacement field (as in the classical continuum of Cauchy), additional variable fields are required to represent the configuration of the body. The continuum theory of Cauchy assumes that the body can be divided into arbitrarily small sub-bodies which, on the one hand, retain the properties of the continuum, and, on the other hand, are sufficiently small so that their configurations are described by the positions of their centres of mass and their dynamics are like that of a Newtonian particle. The mass-centre velocities, when smoothed out, form the macroscopic velocity field of the continuum. However, a body may have microstructure, that is, the character of the body changes when the sub-bodies become smaller than a minimum size; thus the smoothing process must take place at a level of the minimal sub-body that may be too large to be regarded as a simple particle. Instead, this minimal sub-body may be regarded as a microcontinuum with a velocity distribution. The method of virtual power developed by Germain (1973) and others may then be used to determine the force system on the body as a whole by integrating the power of the forces acting on the velocity field within the microcontinuum (Lubliner, 1984). If the velocity distribution in the microcontinuum is determined by derivatives of the macroscopic velocity field, then the resulting continuum model is called a simple multipolar continuum (dipolar if only first derivatives appear). Otherwise, that is, if the velocity distribution in the microcontinuum is determined by parameters independent of the macroscopic velocity field, the result is a generalized continuum model.

The mechanics of elastic media with microstructure was analyzed by Mindlin (1964). He introduced the concept of a unit cell (essentially the same notion as that of the aforementioned microcontinuum), which can be interpreted as a molecule of a polymer, a crystal lattice, or a grain of a granular material. Green and Rivlin (1964) developed the theory of multipolar continuum mechanics, of which Mindlin's theory may be considered a particular case.

In this discussion we will introduce two additional variables that are not included in the traditional formulations of continuum mechanics: porosity and aggregate content. This formulation has been called 'volume-fraction theory', because it introduces the volume fraction of the material as the additional independent kinematic variable. Since the model is based on rational thermomechanics, it will produce an entropy inequality which, when associated with Coleman's method, will permit the derivation of constitutive equations in a natural way. This theory is based on the concept of a generalized continuum within the formal framework of continuum mechanics, which was first used by Goodman and Cowin (1972) for granular materials.

Let B_t denote the configuration of the distributed body at time t, and V_t and M_t the distributed volume and distributed mass respectively. If V_t is absolutely continuous with respect to the Lebesgue volume measure V, then by the Radon-Nikodym theorem there exists a real-valued Lebesgue integrable function $\nu\,(x_i,\,t)$ defined on B_t such that

$$V_t = \int_{P_t} \nu \, dV$$

ν is called the volume distribution function, and is related to the porosity n or the void ratio e by $\nu = 1 - n = 1/(1 + e)$.

Using the same arguments as above, we define an integrable function γ on B_t such that

$$M_t(P_t) = \int_{P_t} \gamma dV_t$$

The function γ is called the distributed mass density. The mass $M_t(P_t)$ can also be expressed as

$$M_t(P_t) = \int_{P_t} \gamma \nu dV$$

where ρ ($= \gamma \nu$) is the classical mass density function.

An extension for concrete (three-phase systems: granules, matrix, and voids) is easily made. Let us decompose the volume distribution ν into ν^g and ν^m with the superscripts referring to the granules and matrix, respectively:

$$\nu = \nu^g + \nu^m$$

The mass densities of the granules and matrix will be denoted ρ^{g*} and ρ^{m*} respectively. The masses of the granules and matrix in a porous granular body P_t are given by

$$M^g = \int_{P_t} \nu^g \rho^{g*} dV$$

$$M^m = \int_{P_t} \nu^m \rho^{m*} dV$$

The total mass of P_t is given by

$$M = M^g + M^m$$

It is convenient to define

$$\rho^g = \nu^g \rho^{g*}$$

and

$$\rho^m = \nu^m \rho^{m*}$$

which can be interpreted as the bulk densities of the granules and the matrix, respectively. The total bulk density of the material is given by

$$\rho = \rho^g + \rho^m$$

Let the following state variables be defined on B_t: the stress tensor T_{ij}, body force b_i, specific energy e, heat flux vector q_i, body heat supply r, specific entropy η, entropy flux vector Φ_t, temperature Θ, equilibrated inertia k, equilibrated stress vector h_i, external equilibrated body force l and intrinsic equilibrated body force g.

Postulating that each of the fields v_i, T_{ij}, b_i, p'_i, k, h_i, l, g', e, q_i, r, and Θ is defined for the matrix and for the granules. All the balance equations, are taken to be valid separately for each phase. The entropy inequality, however, is assumed for the material as a whole.

Conservation of mass
Granules

$$\frac{\partial \rho^g}{\partial t} + \nabla.(\rho^g v^g) = 0$$

Matrix

$$\frac{\partial \rho^m}{\partial t} + \nabla.(\rho^m v^m) = 0$$

v_i is the velocity vector

Balance of linear momentum
Granules

$$\rho^g \frac{dv_i^g}{dt} = t_{ji,j}^g + \rho^g b_i^g + p'^g_l$$

Matrix

$$\rho^m \frac{dv_i^m}{dt} = t_{ji,j}^m + \rho^m b_i^m + p'^m_l$$

Balance of angular momentum
Granules

$$T_{ij}^g = T_{ij}^g$$

Matrix

$$T_{ij}^m = T_{ij}^m$$

Balance of equilibrated force
Granules

$$\rho^g k^g \frac{d\dot{v}^g}{dt} = h_{l,i}^g + \rho^g l^g + g'^g$$

Matrix

$$\rho^m k^g \frac{d\dot{v}^m}{dt} = h_{l,i}^m + \rho^m l^m + g'^m$$

Conservation of energy
Granules

$$\rho^g \dot{e}^g = t_{kl}^g v_{l,k}^g + h_k^g \dot{v}_{,k}^g + q_{k,k}^g + \rho^g r^g - g'^g \dot{v}^g - p_k'^g v_k^g$$

Matrix

$$\rho^m \dot{e}^m = t_{kl}^m v_{l,k}^m + h_k^m \dot{v}_{,k}^m + q_{k,k}^m + \rho^m r^m - g'^m \dot{v}^m - p_k'^m v_k^m$$

Entropy inequality

$$\rho^m \Theta^m \dot{v}^m + \rho^g \Theta^g \dot{v}^g - \Theta^m \left(\frac{q_k^m}{\Theta^m}\right)_{,k} - \Theta^g \left(\frac{q_k^g}{\Theta^g}\right)_{,k} - \rho^m r^m - \rho^g r^g \geq 0$$

The balance of equilibrated force can be considered as a special case of an equation that arises for materials with microstructure. Cowin and Nunziato (1983) pointed out that, with the microstructure theory as framework, the terms g'and $h_{l,i}$ of the balance of equilibrated force can be identified with singularities in the classical linear elasticity known as double force systems without moments. We may combine three double forces without moment along three mutually perpendicular axes and the singularity may be described as a 'centre of compression' or 'centre of dilatation'. Both g'and $h_{l,i}$ can be related to centres of dilatation, while the equilibrated stress vector hi may be related to a single double force system without moment.

The free energies are defined by

$$\Psi^g = e^g - \eta^g \Theta^g$$
$$\Psi^m = e^m - \eta^m \Theta^m$$

Combining these with the entropy inequality yields:

$$-\rho^m(\dot{\Psi}^m + \eta^m \dot{\Theta}^m) + q_k^m/\Theta^m + t_{kl}^m d_{lk}^m + h_k^m \dot{v}_{,k}^m + p_k'^m v_k^m - - \rho^g(\dot{\Psi}^g + \eta^g \dot{\Theta}^g) + q_k^g/\Theta^g + t_{kl}^g d_{lk}^g + h_k^g \dot{v}_{,k}^g +$$

The derivation is fairly general; it neglects only the energy interaction between the granules and the matrix. To specialize this theory to concrete, modeled adequate constitutive postulates should be given. Monteiro and Lubliner (1989) analyzed general models and some of the results are summarized below.

$$\eta^m = -\frac{\partial \Psi^m}{\partial \Theta^m} \qquad \eta^g = -\frac{\partial \Psi^g}{\partial \Theta^g}$$

$$\frac{\partial \Psi^m}{\partial \Theta_{,i}^m} = \frac{\partial \Psi^m}{\partial \dot{C}_{ji}^m} = \frac{\partial \Psi^m}{\partial \dot{v}^m} = \frac{\partial \Psi^m}{\partial v_i^m} = 0$$

$$\frac{\partial \Psi^g}{\partial \Theta_{,i}^g} = \frac{\partial \Psi^g}{\partial \dot{C}_{ji}^g} = \frac{\partial \Psi^g}{\partial \dot{v}^g} = \frac{\partial \Psi^g}{\partial v_i^g} = 0$$

$$g'^m = -\rho^m \frac{\partial \Psi^m}{\partial v^m} \qquad g'^g = -\rho^g \frac{\partial \Psi^g}{\partial v^g}$$

The generalized continuum model is capable of incorporating some of the characteristics of the transition zone, namely its porosity gradient. Work is needed now to use the existing microstructural information obtained by image analysis and check the prediction capacities of the generalized continuum model.

4.4 Conclusions

The mechanical modelling of the transition zone is lagging behind technological developments where the increased understanding and improvement of the transition zone have led to great advances in concrete performance.

Comparison between the Hashin-Strikman bounds for two-phase composites and experimental data for mortar clearly indicates that, for higher sand concentrations, mortar is not a two-phase material even in the elastic domain.

New generalized continuum models are able to incorporate the gradients of porosity existing in the matrix. However, further developments are needed to fully capture the complex microstructure existing in concrete.

Acknowledgments

The author wishes to thank Christine Human for insightful comments on the composite nature of concrete. The financial support given by the Presidential Young Investigator award from NSF is acknowledged.

4.5 References

Cowin, S.C. and Nunziato, J.W. (1983) *Journal of Elasticity*, Vol. 13, pp. 125ff.

Dougill, J.W. (1962) *Journal of the American Concrete Institute, Proceedings*, Vol. 59, pp. 1363ff.

Germain, C.R. (1973) *Academie des Sciences. Comptes Rendus, Paris*, 274, A1051–54.

Goodman, M.A. and Cowin, S.C. (1972) *Arch. Rational Mech. Anal.*, Vol. 44, pp. 249ff.

Green, A.E. and Rivlin R.S. (1964) *Arch. Rational Mech. Anal.*, 17.

Hashin, Z. and Strikman, S., (1963) *Journal of the Mechanics and Physics of Solids*, Vol. 11, pp. 127ff.

Hirsch, T.J. (1962) Microcracking of plain concrete and the shape of the stress–strain curve, *Journal of the American Concrete Institute, Proceedings*, Vol. 59, pp. 427ff.

Hsu, T.T.C., Slate, F.O., Sturman, G.M. and Winter, G. (1963) Microcracking of plain concrete and the shape of the stress-strain curve *Journal of the American Concrete Institute, Proceedings*, Vol. 60, No. 2, pp. 209-24.

Lubliner, J. (1984) Thermomechanics of Deformable Bodies. Class Notes, University of California, Berkeley.

Mindlin, R.D. (1964) *Arch. Rational Mech., and Anal.*, Vol. 16, pp. 51ff.

Monteiro, P.J.M. (1991) *Cement and Concrete Research*, Vol. 21, pp. 947ff.

Monteiro, P.J.M. and Lubliner, J. (1989), *Cement and Concrete Research*, Vol 19, pp. 929.

Nilsen, A.U. and Monteiro, P.J.M. (1993) Concrete: a three phase material, *Cement and Concrete Research*, Vol. 23, pp. 147ff.

Zimmerman, R.W., King, M.S. and Monteiro, P.J.M. (1986) *Cement and Concrete Research*, Vol. 16, pp. 239ff.

5

Micromechanics of the interface in fibre-reinforced cement materials

H. Stang and S.P. Shah

5.1 Introduction

In fibre-reinforced brittle matrix composites the mechanical behaviour of the interface between the fibres and the matrix has a very significant influence on the overall mechanical behaviour of the composite material.

Since brittle matrix composites are designed primarily with the aim of improving the strength and ductility of the brittle matrix material rather than changing the overall stiffness of the matrix material, the ability of the fibres to interact with the toughening mechanisms governing the mechanical behaviour of the matrix material is essential.

Improved ductility and toughness of the matrix material can only be achieved through stabilization or toughening of the matrix cracking process. A number of different stabilization or toughening mechanisms have been identified in the literature including:

- crack blunting (see e.g. Cook and Gordon, 1964),
- crack path deviation (see e.g. Bentur *et al.*, 1985a),
- crack bridging (see e.g. Korczynskyj *et al.*, 1981; Hannant *et al.*, 1983; Selvadurai, 1983; Mori and Mura, 1984; Budiansky *et al.*, 1986; Stang, 1987; Budiansky and Amazigo, 1989),
- crack shielding by microcracking (see e.g. Hutchinson, 1987).

Since matrix cracking in a fibre-reinforced material can only take place with simultaneous interfacial debonding of the interface it is clear that in a description of the mechanical properties of the composite material special emphasis should be put on the mechanical and strength properties of the interface.

The mechanical interfacial properties describe to what extent the fibres can be activated as crack stoppers and crack bridges during the micro- and macrocracking process in the matrix material. Furthermore, since it is the interfacial properties which are responsible for the loading of the fibres, it is also the interfacial properties which determine whether for a given loading of the composite material the fibres are overloaded or broken. Thus, the mechanical properties of the interface together with the mechanical properties of the fibres are the key parameters to consider when the overall strength and ductility of the brittle matrix composite material is to be described.

It has already been demonstrated by theoreticians and experimentalists working in the fields of composite materials, material modelling, and damage mechanics that material behaviour

Interfacial Transition Zone in Concrete. Edited by J.C. Maso. RILEM Report 11.
Published in 1996 by E & FN Spon, 2-6 Boundary Row, London SE1 8HN. ISBN 0 419 20010 X.

can be modelled and predicted given:

- a proper model of the composite material,
- a proper description of each of the material phases included in the composite,
- a proper description of the mechanical interactions on the interface between the material phases, (see e.g. Aveston *et al.*, 1971; Beneviste, 1984; Beneviste and Aboudi, 1984; Beneviste, 1985; Aboudi, 1987; Dollar and Steif, 1988; Aboudi, 1989; Chen and Hui, 1990; Pagano and Tandon, 1990).

In this way valuable insights into the micro- as well as the macromechanics of materials can be gained given a correct modelling of material behaviour on the micro level.

An abundance of different models can be found in the literature describing the overall behaviour of a brittle matrix composite material as a function of the matrix cracking and/or fibre/matrix debonding. These models usually consider the parameters describing the interfacial characteristics as curve fitting parameters which can be used to produce the wanted overall behaviour of the composite material.

On the other hand, much work has been done in order to be able to characterize experimentally the mechanical behaviour of the fibre–matrix interface using, for example, fibre pull-out or fibre indentation tests. These experiments are often interpreted with simple models which are often conceptually different from the continuum mechanical models used when the behaviour of the composite material is modelled. As a consequence of this it is often not possible to transfer the parameters determined experimentally to the continuum mechanical models for the composite material behaviour.

With respect to brittle matrix composite materials and especially cementitious matrix composite materials there is no well established way of characterizing the fibre–matrix interface on the micro level either from a theoretical or an experimental point of view.

The purpose of the present study is to summarize the different approaches suggested in the literature, to point out advantages and disadvantages and in this way encourage the development of both experimentally and theoretically well-founded models for the interface in cementitious composite materials.

In reviewing the literature, brittle matrix composites in general have been covered, that is, the review has not been confined to literature concerning cementitious composites. Some models which were originally designed for metal–matrix composites have even been included. In this way the risk of missing important sources of inspiration has been minimized. In this respect it should be noted that even though a composite material has a ductile matrix the fibre–matrix interfacial behaviour can still be similar to the interfacial behaviour of a brittle matrix composite material since it is generally accepted that the mechanical behaviour of the interfaces in a given composite material system have to be regarded totally independently from the mechanical behaviour of the bulk matrix and fibre material.

The present study first of all investigates the geometrical modelling of the interface used by different models in order to present a precise definition of the term 'interface'. Two basic approaches are used in the literature, a volumetric and a surface-oriented definition. Basically not much is gained from a volumetric description of the interface and in the following sections we concentrate on surface-based interfacial descriptions.

Secondly, the constitutive description of the interface is investigated. The constitutive relations of interfaces are different from the constitutive relations of the bulk matrix and fibre material since they involve not stress and strain but stress and displacement. Often the interface is divided into a perfectly bonded and a debonded part with two separate sets of constitutive relations. However, so-called cohesive models which do not distinguish between 'bonded' and 'debonded' have been suggested for metal–matrix composite materials. The characteristics of these models are described and it is shown that certain models suggested for cementitious materials can be considered as special cases of these cohesive models.

If the interface is divided into a perfectly bonded and a debonded part a criterion must be set up to govern the transition of bonded interface to debonded interface. Basically, two types of debonding criteria can be found in the literature: a strength-based and a fracture-mechanics based criterion. The two types of criteria are compared and discussed in connection with an evaluation of the modelling of the fibre pull-out problem.

It should be noted that much work has been done with respect to characterization of the nature of linear elastic stress and strain fields near the tip of bi-material (interface) cracks. However, the present study does not go into details about the nature of the crack tip fields at the interface, in this study it is sufficient to realize that a linear elastic analysis predicts a singularity at the crack tip.

5.2 Geometrical characterization of the interface

The first step in a modelling process is to specify the geometrical characteristics of the phenomenon to be investigated. In this process a notation is usually defined. In the case of interfaces in cementitious composites there is not complete agreement in the literature regarding the notation to be used in interfacial models.

Frequently the word 'interface' is related to the *surface* separating the different material phases in the composite material.

A transition zone between the bulk matrix material and the fibre is a well known phenomenon in many kinds of composite material systems including cementitious composite materials, see Pinchin and Tabor, 1978a; Barnes *et al.*, 1978; Page, 1982; Bentur *et al.*, 1985b; and Wei *et al.*, 1986. The transition zone is typically more porous and less stiff than the bulk matrix material and there is thus good reason for this zone to be included in the theoretical analysis of the composite material. However, in some studies the transition zone is referred to as the 'interface' or the 'interfacial zone'.

In other models the word 'interface' is used to designate a volume rather than a surface. This is typically the case in shear lag types of analysis, see e.g. Palley and Stevans (1989), where the shear lag which is responsible for the total shear deformation is denoted the 'interface'. Another model of the same type is the model of Lhotellier and Brinson (1988) where a double shear lag is introduced around the fibre. In this model the outer shear lag is denoted the 'matrix' while the inner is denoted the 'interface'.

In the present study the word 'interface' will always refer to the *surface* in 3D analysis and the *line* or *curve* in 2D analysis which separates the matrix material (with or without transition zone) from the fibre material, while the 'transition zone' always refers to a volume. See Fig. 5.1.

Furthermore, the present study is only concerned with the constitutive relations for stresses and displacements on the interface. Probably, the presence of a transition zone can, in many cases, be incorporated in the constitutive relations for the interface as outlined in Fig. 5.2 which shows the two different ways of including the transition zone in the analysis. Either the transition zone is explicitly incorporated in the model as a volume with elastic moduli which differs from the elastic moduli of the matrix material, or the presence of the transition zone is incorporated in the constitutive relation for the interface. Modelling the transition

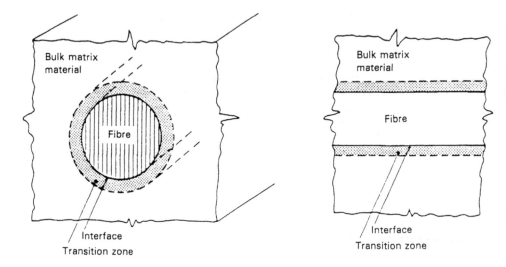

Fig. 5.1 Terminology used in the present study: the surface in 3D analysis (left) and the line in 2D analysis (right) separating the fibre from the matrix is called the *interface* or fibre-matrix interface. A possible *transition zone* is a volume (3D) or area (2D) of matrix material surrounding the fibre. This matrix material is typically more porous and thus less stiff than the bulk matrix material.

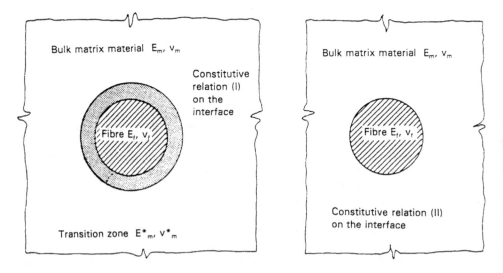

Fig. 5.2 Two ways of modelling the transition zone. Either the transition zone is explicitly incorporated in the model as a volume with elastic moduli which differ from the elastic moduli of the matrix material, or the presence of the transition zone is incorporated in the constitutive relation for the interface. Thus, the constitutive relations for the interface in the two models outlined above or the parameters in a given constitutive relation should in general be different as indicated in the figure.

zone with a special constitutive relationship for the interface is exactly the procedure suggested by Steif and Hoysan (1986), who show that an interfacial shear stiffness relating the displacement jump and the shear stress can be adjusted to accommodate the presence of either a transition zone or an interfacial crack.

Eliminating the transition zone from the analytical model of course simplifies the model a great deal and furthermore removes any ambiguity regarding what value of thickness and stiffness of the transition zone to use in the analytical modelling. In this connection it should be noted that a direct experimental determination of the properties (strength and stiffness) of the transition zone is extremely difficult if not impossible. The properties of the transition zone can only be determined indirectly through e.g. microhardness measurements (see e.g. Wei *et al.*, 1986)

A geometrical transition zone is explicitly incorporated in the work of Lhotellier and Brinson (1988) in their two-layer shear lag model though their aim primarily is to describe fibre-reinforced composites with polymer matrices. Furthermore, any shear lag model, e.g. Palley and Stevans (1989), and Stang and Shah (1990), can be interpreted as a model describing a fibre surrounded by a transition zone since the shear lag is given a description which is detached from the description of the bulk matrix material.

5.3 Mechanical characterization of the interface

From a mechanical point of view the interface in cementitious composite materials can be characterized using a variety of different principles.

The different principles found in the literature can roughly be divided into three main groups:

- models describing a perfectly bonded interface,
- models describing a debonded interface,
- models describing cohesive or the cohesively bonded interfaces.

Perfectly bonded interfaces can be interpreted physically as interfaces where a chemical and/or physical bond has been established between the fibre material and the matrix material. This bond is strong enough to ensure both stress and displacement continuity across the interface given an external load on the system.

Debonded interfaces can be interpreted as a previously bonded interface which has been loaded beyond the strength of the bonded interface. The constitutive relations on the debonded interface primarily relate to the kinematics of the interface i.e. they describe frictional and contact phenomena.

Cohesive interface models are interface models which assume that stress transfer across the interface is always connected with relative fibre–matrix displacement. Thus it is assumed that no physical and/or chemical adhesive bonds between the fibre and the matrix exist at any time. In this case stress transfer is directly connected to friction and physical mismatch effects, and only relative displacements can activate a stress transfer, see Fig. 5.3.

Cohesive interface models have been used successfully in plastically deforming solids and composite materials, (Needleman, 1987; Nutt and Needleman, 1987; Needleman and Nutt, 1989; Tvergaard, 1989; Needleman, 1990a, 1990b, 1990c) but have also been used in simple fibre pull-out models in cementitious materials, (Wang *et al.*, 1988, 1989; and Nammur and Naaman, 1989).

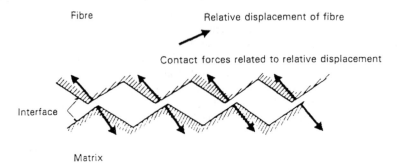

Fig. 5.3 Surface roughness and physical mismatch in connection with displacement discontinuities are assumed to be responsible for stress transfer in debonded and cohesive interface models.

The different interface models are described in detail in the next sections where the following notation is adopted. All quantities related to the fibre material are marked with subscript *f* while all quantities related to the matrix material are marked with a subscript *m*.

It is assumed that only axial symmetric configurations are considered, thus even 3D descriptions can be reduced to the 2D description shown in Fig. 5.4 where the sign convention used in this description is also outlined.

Some types of analysis, such as the shear lag types, are simplified so that not all the stress and displacement components shown in Fig. 5.4 are actually part of the analysis. However, the quantities should be considered as a general framework suitable for the description of different types of analysis.

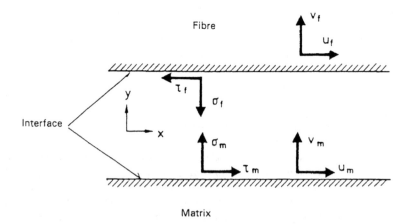

Fig. 5.4 The notation and the sign convention used in the present description of constitutive relations on the fibre-matrix interface. The fibre and the matrix are moved apart for clarity. The displacements are denoted *u* and *v* for the in-plane and normal component, respectively while the interface stresses are denoted *τ* and *σ* for the shear and the normal component, respectively.

5.3.1 THE PERFECTLY BONDED INTERFACE

As mentioned above the perfectly bonded interface requires displacement continuity as well as interfacial stress (traction) continuity. This requirement is simply expressed as

$$
\left.
\begin{aligned}
\tau_f &= \tau_m \\
\sigma_f &= \sigma_m \\
u_f &= u_m \\
v_f &= v_m
\end{aligned}
\right\} \quad \text{on } I_b
\tag{5.1}
$$

where I_d represents the perfectly bonded part of the fibre–matrix interface.

Many simplified types of analysis (typically shear lag models) involve only the determination of in-plane displacement and shear stresses on the interface, thus in these models the perfectly bonded interface is characterized by the following equations

$$
\left.
\begin{aligned}
\tau_f &= \tau_m \\
u_f &= u_m
\end{aligned}
\right\} \quad \text{on } I_b
\tag{5.2}
$$

The greater part of the models found in the literature use a description of a perfectly bonded interface corresponding to the equations above. As mentioned above the description implies that an *adhesive* bond exists between the fibre and the matrix. This bond can be of a physical and/or chemical nature.

The adhesive bond is often denoted the *elastic* bond and the term *elastically bonded interface* is sometimes used to describe the perfectly bonded interface.

If the interface is assumed to be characterized by a finite strength (in cementitious materials the strength of the interfaces are usually considerably lower than the strengths of both the cementitious matrix and the fibre) then it is necessary to be able to include a strength criterion in the description of the perfectly bonded interface. This criterion—which typically is local i.e. it is related to the stress state at a point—then predicts when a point on the bonded interface changes state and becomes a point on a debonded interface.

The problem of determining criteria for debonding of perfectly bonded interfaces is treated in a following section.

5.3.2 THE DEBONDED INTERFACE

On the debonded interface the characterization is changed from a continuity condition into a prescribed surface traction boundary condition which is applied to both the fibre surface and the matrix surface in the following way:

$$
\left.
\begin{aligned}
\tau_f &= \tau_m = f \\
u_f &= u_m = g
\end{aligned}
\right\} \quad \text{on } I_d
\tag{5.3}
$$

where I_d represent all points on the debonded interfacial surface while f and g represent the prescribed surface traction in the general case.

This description obviously allows for displacement discontinuities along the debonded interface, however, it does not give any guarantee that surface overlapping will not occur. Thus equation (5.3) is only sufficient if the solution shows that:

$$v_f - v_m \geq 0 \quad on \; I_d \tag{5.4}$$

If equations (5.3) and (5.4) are not satisfied at all points on the debonded interface the nature of the problem changes and even a linear elastic analysis becomes non-linear since a contact problem is involved. In this case a complete analysis becomes very complicated since the mechanical behaviour of the interface is governed by either the equations (5.3) or by a mixed type of boundary condition which is described in the following. Furthermore, it is generally not possible in advance to determine which part of the debonded interface is governed by equations (5.3) and which part is governed by the mixed type of boundary condition. However, before looking at the mixed type of boundary conditions some simplifications of relation (5.3) will be considered.

The surface reactions f and g represent frictional and/or cohesive stresses. These frictional stresses can be assumed to be vanishing so that equation 5.3 reads:

$$\left.\begin{array}{l} \tau_f = \tau_m = 0 \\ u_f = u_m = 0 \end{array}\right\} \quad on \; I_d \tag{5.5}$$

This corresponds to the stress-free surface description known from classic crack descriptions, see e.g. Atkinson *et al.*, 1982; Budiansky *et al.*, 1986; Stang and Shah, 1986; and Morrison *et al.*, 1988.

However, it is argued by many authors that frictional and/or cohesive interface stresses play an important role in many fibre–cementitious matrix composite systems, especially systems with relatively stiff and hard fibre types, see e.g. Gao *et al.* (1988). In fact, it is argued in Wang *et al.* (1988) that in most cases only the debonded region of the interface of the fibre in a fibre pull-out test and the frictional shear stresses on this part of the interface needs to be included in the modelling while the perfectly bonded region of the interface can be disregarded.

Thus, often the frictional surface tractions are assumed to be non-vanishing. The simplest assumption – that the surface tractions are constant – reduces equations (5.3) to:

$$\left.\begin{array}{l} \tau_f = \tau_m = \tau* \\ \sigma_f = \sigma_m = 0 \end{array}\right\} \quad on \; I_d \tag{5.6}$$

See Lawrence, 1972; Laws *et al.*, 1973; Bartos, 1981; Laws, 1982; Gray, 1984; Gopalaratman and Shah, 1987; Palley and Stevans, 1989; and Stang, Li, and Shah, 1990.

In simplified shear lag analysis only the shear stresses are determined, thus in this type of analysis the debonded interface is characterized by:

$$\tau_f = \tau_m = f \quad on \; I_d \tag{5.7}$$

which in the two simplifying cases outlined above reads:

$$\tau_f = \tau_m = 0 \quad on \; I_d \tag{5.8}$$

and

$$\tau_f = \tau_m = \tau * \qquad \text{on } I_d \tag{5.9}$$

Some models, (Pinchin and Tabor, 1978b, 1978c; Beaumont and Aleszka, 1978; Wells and Beaumont, 1985; Budiansky *et al.*, 1986; Gao, 1987; Abudi, 1989; Sigl and Evans, 1989; Hsueh, 1990a, 1990b, 1990c) deal specifically with the case where the matrix exerts a *compressive* stress on the debonded interface due to thermal mismatch, mechanical loading, or matrix shrinkage. As mentioned above, the boundary condition on the debonded interface in that case becomes a complicated mixed type of boundary condition requiring displacement continuity perpendicular to the interface:

$$v_f = v_m \qquad \text{on } I_d \tag{5.10}$$

along with stress continuity *perpendicular* to the interface:

$$\sigma_f = \sigma_m \qquad \text{on } I_d \tag{5.11}$$

However, the surface traction *in* the interface plane is a prescribed frictional surface traction. This condition can be written as:

$$|\tau_f| = |\tau_m| = f \qquad \text{on } I_d \tag{5.12}$$

The magnitude of the prescribed frictional surface traction is usually assumed to depend on the compressive (negative) stress normal to the interface through

$$f = -\mu\sigma_f \quad (-\sigma_f \geq 0) \tag{5.13}$$

where μ is the frictional coefficient, (Pinchin and Tabor, 1978b, 1978c; Beaumont and Aleszka, 1978; Wells and Beaumont, 1985; Budiansky *et al.*, 1986; Gao, 1987; Sigl and Evans, 1989), or the friction can be assumed to depend on the normal stresses on the interface according to the Coulomb frictional law (Aboudi, 1989):

$$f = c - \mu\sigma_f \quad (c - \mu\sigma_f) \geq 0 \tag{5.14}$$

where c is a measure for the cohesion.

In the simplified shear lag analysis of e.g. Pinchin and Tabor (1978b, 1978c), Beaumont and Aleszka (1978), and in the solution referred to in Wells and Beaumont (1985) the displacements perpendicular to the interface do not enter the analysis. Thus, in these cases the following relation on the interface is used:

$$\left. \begin{array}{l} \sigma_f = \sigma_m \\ |\tau_f| = \tau_m| = f \\ f = -\mu\sigma_f \quad -\sigma_f \geq 0 \end{array} \right\} \qquad \text{on } I_d \tag{5.15}$$

where the contact stresses σ_f and σ_m are related to the volume changes (shrinkage, thermal mismatch), external load and Poisson's ratio for the matrix and the fibre material independently of the shear lag solution.

The effect of Poisson's ratio on the contact stresses and thus on the frictional stresses are investigated in detail by Hsueh (1990a, 1990b, 1990c) who investigated both fibre pull-out and fibre push-down (indentation) theoretically and experimentally in a fibre-reinforced ceramic composite.

As indicated in equations (5.13), (5.14) and (5.15) it is usually assumed that frictional stresses are only active when the resulting normal stress in the interface is negative (compressive contact, equation (5.14)) or at least less that a critical value (equation (5.15)) with a critical value of c/μ) while it is assumed that the frictional stresses are eliminated in the case of a (large enough) normal displacement discontinuity – expressed in terms of the normal stresses.

The direction of the frictional surface traction is evident as long as the loading and the direction of sliding is monotonic in time and in fact all the above relations are valid under the implicit assumption that

$$\frac{du_d}{dt} = \dot{u}_d > 0 \tag{5.16}$$

with

$$u_d = u_f - u_m \tag{5.17}$$

However, in more general cases the direction of the frictional stress might change. It can be assumed that the direction of the shear stress is determined from the direction of the displacement discontinuity *rate*, as suggested by e.g. Wang, Li and Backer (1988) for a cementitious matrix composite material and by Abudi (1989) for a more general composite material system. The suggested relation between displacement discontinuity rate and direction of the frictional stress is given by:

$$sgn(f) = sgn(\dot{u}_d) \tag{5.18}$$

where *sgn(f)* refers to the sign of *f*.

A relation for *f* which introduces both the Coulomb frictional law and at the same time ensures the correct direction for τ (according to the above assumption) and distinguishes between contact and non-contact was introduced by Abudi (1989):

$$f = sgn(\dot{u}_d) \, H \, (c - \mu\sigma_f) \, (c - \mu\sigma_f) \tag{5.19}$$

where *H* denotes Heaviside's function, given by

$$H(x) = \begin{cases} 0 \text{ for } x < 0 \\ 1 \text{ for } x > 0 \end{cases} \tag{5.20}$$

Note that the type of description used by Abudi implies that tensile stresses can be transferred across the *debonded* interface. Though this is in accordance with many crack descriptions used in cementitious materials (see e.g. Hillerborg *et al.* (1976) and many others) one would expect a relationship between tensile stress transfer and the magnitude of the normal displacement discontinuity. No such relation enters any of the models mentioned above. In fact the model of Abudi does not allow any normal displacement discontinuity along the debonded interface whether the normal stress state is tension or compression. This of course simplifies the analysis but it is in direct contrast to many experimental observations on cementitious composites, (Bentur *et al.*, 1985a, 1985b).

5.3.3 THE COHESIVE INTERFACE

In the cohesive interface models, it is assumed that only relative displacements between the fibre and the matrix can activate stress transfer at the interface. Furthermore, the interface traction is described as a function of the displacement discontinuity, thus allowing for introduction of the description of normal tensile stress degradation with increasing normal displacement discontinuity, the description missing in the debonded interface models treated in the previous section.

Since there is a unique relationship between interface traction and interface displacement discontinuity at all times, it is not necessary to distinguish between the bonded and debonded part of the interface.

If one uses the presence of displacement discontinuities on the interface to characterize a debonded interface one can say that in the cohesive interface models it is assumed that the interface is debonded for any interface traction.

Alternatively, the cohesive interface models can be described as a perfectly bonded–debonded interface model where the strength of the bonded interface is vanishing and where the model on the debonded part of the interface can be described as a relation between interface traction and displacement discontinuity.

Consequently, the cohesive interface model can be interpreted physically as an interface with no adhesion but with an interface traction which is connected to surface roughness in such a way that a relative displacement between the fibre surface and the matrix surface causes interface tractions to develop.

In general, a cohesive interface model can be described in the following way:

$$\left.\begin{array}{llll} f = f(u_d, v_d) & \text{for} & u_d \neq 0 & \vee & v_d \neq 0 \\ g = g(u_d, v_d) & \text{for} & u_d \neq 0 & \vee & v_d \neq 0 \\ f = 0 & \text{for} & u_d = 0 & \wedge & v_d = 0 \\ g = 0 & \text{for} & u_d = 0 & \wedge & v_d = 0 \end{array}\right\} \quad \text{on } I \qquad (5.21)$$

where I denotes all points on the interface and where

$$u_d = u_f - u_m \qquad (5.22)$$

and

$$v_d = v_f - v_m \qquad (5.23)$$

The cohesive model in its simplest form can be found in Wang *et al.* (1989) and in Li (1990) where the constitutive interface relation in a fibre pull-out model is described as

where I denotes the total interfacial surface and where f is assumed to be a positive constant. As indicated, only $u_d \geq 0$ is considered here.

A more refined version of equation (24) which is able to model loading and subsequent

$$\left.\begin{array}{l} \tau_f = \tau_m = f \ \text{for} \ \ u_d > 0 \\ \tau_f = \tau_m = 0 \ \ \text{for} \ \ u_d = 0 \end{array}\right\} \quad \text{on } I \tag{5.24}$$

unloading, was introduced in a fibre pull-out model by Wang *et al.* (1988). Here the frictional stresses were introduced as functions of the magnitude of the relative slip and the rate of change of the relative slip so that:

$$\left.\begin{array}{l} |\tau_f| = |\tau_m| = f(u_d) \ \text{for} \ \ u_d > 0 \\ \tau_f = \tau_m = 0 \ \ \ \ \text{for} \ \ u_d = 0 \\ sgn(\tau) = sgn(\dot{u}_d) \end{array}\right\} \quad \text{on } I \tag{5.25}$$

In these models a parabolic relationship between f and the displacement discontinuity was assumed and both slip-hardening as well as slip-softening could be modelled. See Fig. 5.5. It was found that a slip-softening model describes the pull-out of a steel fibre from a cementitious matrix well, while the slip-hardening model proved to be more suitable for the description of pull-out of a synthetic fibre from a cementitious matrix. The physical interpretation of slip-softening is, according to Wang *et al.* (1988), that a gradual breakdown of the cement matrix takes place due to the stiffness and the hardness of the steel fibre while slip-hardening can be interpreted as a consequence of fibre surface abrasion and accumulation of wear debris.

Note that the cohesive models described above use a relationship for shear stress and slip which is discontinuous in the points where the displacement discontinuity u_d changes

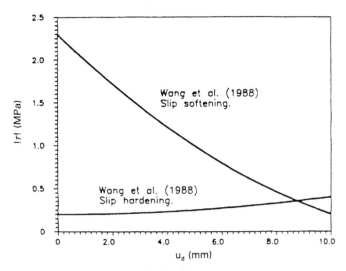

Fig. 5.5 The slip-hardening and slip-softening models of Wang, Li and Backer (1988). The slip-hardening relationship was used in the modelling of the pull-out of synthetic fibres from a cementitious matrix while the slip-softening relationship was used in the modelling of the pull-out of steel fibres from a cementitious matrix.

(continuously) from 0 to a non-zero finite value. See Fig. 5.6, where a continuous cohesive model presented by Nammur and Naaman (1989)is also shown.

It is not clear whether such a discontinuity has any physical interpretation, however it must not be concluded that stresses are transferred on the part of the interface where there is no relative displacement; i.e. the value of f for small u_d *cannot* be interpreted as the strength of the interface with no displacement discontinuity.

In Wang *et al.* (1988) a perfectly bonded interface is simply excluded from the analysis since it is argued that the introduction of a perfectly bonded interface typically has very little significance on the pull-out load-displacement relationship in cementitious composite systems. So in this case the shear stress discontinuity at $u_d = 0$ is simply a consequence of the model assumptions.

As mentioned above, a continuous cohesive interface model was introduced by Nammur and Naaman (1989) using a relationship identical to equation (5.25). Nammur and Naaman used a piece-wise linear shear stress–slip relationship as shown in Fig. 5.6. In the Nammur and Naaman relationship $f \to 0$ for $u_d \to 0$ in agreement with the fact that the strength of the bonded interface has a vanishing strength and thus eliminating any traction discontinuity at the end of the debonded zone.

The primary interest of Nammur and Naaman (1989) is to describe the initial part of the fibre pull-out process and, consequently, slip-hardening or softening effects are not included in their model. A model able to describe the initial as well as the final stages of a fibre pull-out experiment could probably be achieved by combining the Wang *et al.* model with the Nammur and Naaman model.

The most general version of cohesive interface models was introduced by Needleman using a full version of equation (5.21).

The model was originally introduced by Needleman (1987) for use in plastically deforming solids and composite materials and has since been elaborated on and modified in a number of papers, Needleman (1990a, 1990b, 1990c). However, since simplified versions seem to

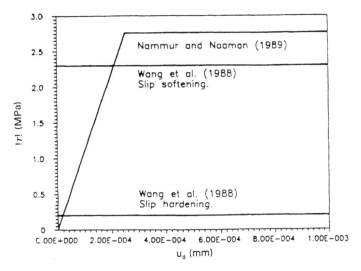

Fig. 5.6 The discontinuous cohesive interface model of Wang *et al.* (1988) shown on a displacement scale where the difference between the continuous model of Nammur and Naaman (1989) is clear.

be working fine in cementitious composites, it seems obvious to use this kind of interface description in brittle matrix composites as well.

The functions f and g are determined using a potential formulation. The potential is specified in terms of three parameters: a maximum tensile surface traction σ^{max} corresponding to a positive interface separation, a maximum positive interface separation v_d^{max} and the ratio between the shear and the normal stiffness of the interface α.

The potential is chosen so that the normal surface traction component reaches a maximum for increasing normal interface separation and then drops to zero when the normal interface separation exceeds v_d^{max}. In the case of negative interface separation (interface overlapping), compressive normal surface tractions build up. The in-plane surface tractions are assumed to depend linearly on the in-plane displacement discontinuity u_d, however, the in-plane tractions also drop to zero when the normal separation exceeds v_d^{max}, See Fig. 5.7.

This model has been used in a couple of papers for studying void formation at inclusion boundaries (Nutt and Needleman, 1987; Needleman and Nutt, 1989). Tvergaard (1989) modified the model in order to include Coulomb friction and used the modified model to study debonding in whisker-reinforced metals.

The model is somewhat similar to the cohesive crack models originally suggested by Barenblatt (1962), and suggested for use in cementitious composite materials such as concrete by Hillerborg, (Hillerborg *et al.* (1976), Hillerborg (1980)). The cohesive interface models are appealing mainly due to the fact that:

- the contact problem of the interface has been removed, i.e. it is possible to use the same boundary problem formulation on the entire interface.

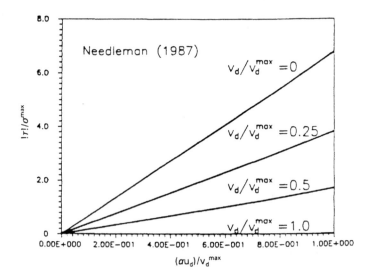

Fig. 5.7 The normalized in-plane surface traction component as a function of normalized in-plane displacement discontinuity as predicted by the Needleman (1987) model. The relationship is specified in terms of three parameters: a maximum tensile surface traction σ^{max} corresponding to a positive interface separation, a maximum positive interface separation v_d^{max}, and the ratio between the shear and the normal stiffness of the interface.

- the interface traction degradation with increasing interface displacement discontinuity which is well-known from other crack descriptions in cementitious materials is included in the formulation.

However, it is clearly a drawback that

- a possible adhesion at the fibre–matrix interface is not included in the analysis. The significance of the fibre–matrix adhesion is discussed in a later section.
- negative normal displacement discontinuities v_d (overlapping) - which from a kinematically point of view are inadmissible - are allowed in the model.

5.4 Debonding criteria

Assume that a model for a fibre embedded in a cementitious matrix has been established including an interface model derived from the general interface description above. If the interface is divided into a perfectly bonded and a debonded interface, a criterion is needed in order to determine whether a point on the perfectly bonded interface is about to change status and becoming a point on the debonded interface.

Basically debonding criteria can be divided into two classes: the stress-based predicting onset of debonding when the interface traction reaches some critical value, and the fracture mechanics approach which interprets the debonded interface as a crack, the bonded interface as the potential crack path, and formulates the criterion for debonding on a fracture mechanics basis.

In the following sections the two approaches will be discussed in more detail and the difference between the two will be examined in the case of fibre pull-out and finally in more general terms in the discussion.

Note that if the interface model is a cohesive interface type of model no additional criterion is needed to predict debonding since the same displacement/interface traction relationship is governing the behaviour of the interface at all times.

5.4.1 STRESS-BASED CRITERIA FOR DEBONDING

Assume that a model for a composite material has been set up which includes a perfectly bonded–debonded interface model.

In general terms a stress-based criterion for debonding can be expressed as a failure criterion in either τ_f and σ_f or τ_m and σ_m on the bonded interface I_b. A general failure criterion can be written as

$$F(\tau, \sigma, p_i) = 0 \quad i = 1, \ldots, n \tag{5.26}$$

where p_i is one of n strength parameters. Due to the traction continuity relation (5.1) it makes no difference whether the failure criterion is expressed in fibre stress or matrix stress, thus τ and σ means either τ_f and σ_f or τ_m and σ_m. The fulfilment of equation (5.26) in a point on the bonded interface correspond to a change of status of that point from "perfectly bonded" to 'debonded', the significance of these terms being determined by the type of model used.

If a traditional shear lag type of analysis is used, equation (5.26) is simplified a great deal since only the shear traction is determined in the analysis. Usually only one strength parameter is used in the failure criterion, (e.g. de Vekey and Majumdar, 1968; Greszczuk, 1969; Lawrence, 1972; Takaku and Arridge, 1973; Bartos, 1980; Laws, 1982; see also the review paper by Gray, 1984), reducing equation (5.26) to

$$|\tau| = \tau^{crit} \tag{5.27}$$

where τ^{crit} is the (constant) strength parameter.

In some models, a Coulomb type of failure criterion has been used, (see e.g. Pinchin and Tabor, 1978b, 1978c; Beaumont and Aleszka, 1978; Aboudi, 1989) extending the criterion (5.27) to a two-parameter failure criterion:

$$|\tau| = c - \mu\sigma \qquad (c - \mu\sigma) \geq 0 \tag{5.28}$$

where c is a measure for the cohesion, and μ is the frictional coefficient.

Strength parameters related to a shear lag type of analysis of the fibre pull-out problem for a number of fibre–matrix systems can be found in Laws (1982) and Bentur and Mindess (1990).

Even though the idea of using a stress-based debonding criterion on the interface seems reasonable from an engineering point of view some difficulties with this approach should be pointed out here.

It is well-known that a complete linear elastic analysis of the interface problem yields stress singularities at the transition point between the perfectly bonded and the debonded interface. These singularities are not captured with a crude analysis like a simple shear lag analysis. But the presence of these singularities is the reason why, for instance, finite element analysis of the interface problem becomes mesh-dependent with respect to the interface stresses, see e.g. Marmonier *et al.* (1988). Even experimentally, stress concentrations near the debonding front in cementitious composites have been observed which are unlikely to be predicted accurately by shear lag analysis, Bowling and Groves (1979).

These observations regarding the interface stresses and their dependence on the type of analysis performed has some serious consequences for the parameters introduced in the stress-based debonding criteria. The parameters cannot be regarded as material parameters in the usual sense since their values vary wildly according to the type and complexity of the analysis used. A given value of the interface strength parameters is connected to a specific type of analysis which of course makes utilization of these parameters difficult. Furthermore, it is likely—due to the incompleteness of the analysis—that the value of the strength parameters corresponding to a given interface will depend on the geometry and the loading conditions on the composite system. Size effects have been observed in fibre pull-out tests, (Naaman and Shah, 1976; Wells and Beaumont, 1982, 1985), which a pull-out analysis using a stress-based debonding criterion fails to predict.

5.4.2 FRACTURE MECHANICS CRITERIA FOR DEBONDING

The other approach to the debonding problem is the fracture mechanics approach which was applied to the analysis of fibre-reinforced brittle matrix composites by e.g. Gurney and Hunt (1967), Outwater and Murphy (1969), Bowling and Groves (1979), Wells and Beaumont

(1982), Piggott *et al.* (1985), Wells and Beaumont (1985), Stang and Shah (1986), Piggott (1987), Gao (1987), Mandel *et al.* (1987), Morrison *et al.* (1988), Gao *et al.* 1988, Sigl and Evans (1989), Hamoush and Salami (1990), Hutchinson and Jensen (1990), Stang *et al.* (1990).

Almost exclusively the fracture mechanics postulate has been related to the total energy release rate G, where

$$G = \frac{\partial W_{ex}}{\partial a} - \frac{\partial W_{\varepsilon}}{\partial a} - \frac{\partial W_f}{\partial a} \tag{5.29}$$

Here a is the projected area of the debonded zone, W_{ex} is work done by prescribed external forces, W_{ε} is elastic strain energy, and W_f is dissipation in inelastic parts of the structure, e.g. work done by friction on the debonded interface, (Marshall *et al.*, 1985; Gao *et al.*, 1988; Sigl and Evans, 1990; Stang *et al.*, 1990).

The just mentioned calculation of the energy release rate is convenient when using an analytical solution e.g. based on a shear lag analysis. Using FEM with special crack tip element modelling the energy release rate can be evaluated as a function of the stress intensity factors, k_I and k_{II}:

$$G = G(k_I^2 + k_{II}^2) \tag{5.30}$$

Mandel *et al.* (1987), Atkinson *et al.* (1982), Stern and Hong (1976), Sih and Rice (1964).

The criterion for debonding is then given by equating the energy release rate with the critical energy release rate G_i^{crit} of the interface:

$$G_i^{crit} = G \tag{5.31}$$

In fibre pull-out situations it is often assumed, implicitly or explicitly, that the critical energy release rate corresponds to the mode II toughness of the interface G_{III}^{crit}, i.e. that the mode I contribution to the energy release rate is insignificant. Thus, with this assumption, equation (5.31) can be restated more precisely as

$$G_{iII}^{crit} = G \tag{5.32}$$

The mode II interface toughness of various fibre–cementitious matrix systems have been determined experimentally using different types of analytical modelling to determine the energy release rate from the fibre pull-out set up, (Stang and Shah, 1986; Mandel *et al.*, 1987; Hamoush and Salami, 1990). It has been shown that, in general, the mode II interface toughness for a given fibre–matrix system is less than the mode I toughness for the matrix alone (Hamoush and Salami, 1990). Furthermore, the interface toughness depends heavily on the microstructure of the matrix (Mandel *et al.*, 1987).

Some experimentally determined values for G_{III}^{crit} from the references (1) Stang and Shah (1986), (2) Hamoush and Salami (1990), and (3) Mandel *et al.* (1987) are shown in Table 5.1. The fibre–matrix systems are (a) smooth steel fibre 0.4 × 13 mm in Portland cement mortar (w/c = 0.55) (Naaman and Shah, 1976), (b) smooth brass coated steel fibre 0.25 × 13 mm in Portland cement mortar (w/c = 0.6) (Naaman and Shah, 1976), (c) smooth steel

Table 5.1. Experimentally determined values for G_{ill}^{crit} Nm/m$_2$ from the references (1) Stang and Shah (1986), (2) Hamoush and Salami (1990), and (3) Mandel *et al.* (1987). A description of the fibre–matrix systems can be found in the text.

Fibre–matrix	Ref. (1)	Ref. (2)	Ref. (3)
(a)	26.6	24.2	
(b)	43.4	41.0	
(c)	11.0	10.6	
(d)	16.6	15.2	
(e)	-	-	27.0
(f)	-	-	23.2
(g)	-	-	29.8
(h)	-	-	30.2
(i)	-	-	6.8
(j)	-	-	7.2
(k)	-	-	4.2
(l)	-	-	4.0

fibre 0.15 × 13 mm in Portland cement mortar (w/c = 0.55) (Naaman and Shah, 1976), (d) smooth steel fibre 0.38 × 30 mm in Portland cement mortar (w/c = 0.55) (Burakiewicz, 1978), (e)-(h) smooth steel fibres 0.5 × 20 mm in polymer-modified cement mortar (w/c = 0.3) fibre spacing 25, 16.3, 12.5, 10 × fibre diameter, (i)-(l) smooth steel fibres 0.5 × 20 mm in cement mortar (w/c = 0.42) fibre spacing 25, 16.3, 12.5, 10 × fibre diameter.

Some work has been done recently (Jensen *et al.*, 1990, Jensen, 1990) in order to separate the mode I and mode II contribution to the energy release rate, thus introducing debonding criteria of the mixed mode type:

$$F(G_I, G_{II}, G_{iI}^{crit}, G_{iII}^{crit}) = 0 \qquad (5.33)$$

where G_I and G_{II} are the contributions to the energy release rate from mode I and mode II opening, respectively, while G_{iI}^{crit} and G_{iII}^{crit} are the mode I and II toughness of the interface, respectively.

Mixed mode criteria for debonding have been used mainly to describe delamination of fibre-reinforced polymer composites, and to the authors' knowledge no attempt has yet been done to introduce mixed mode criteria on interfaces in cementitious and other brittle matrix composites.

It should be noted that the calculation of the energy release rate does not suffer from the difficulties encountered in the calculation of interfacial stresses. Even simple models of the shear lag type are known to produce relatively good approximations for the energy release rate. Thus interpretation of experimental observations using a fracture mechanics debonding criterion in connection with a relatively simple model is likely to produce reliable results for the interfacial toughness, results which can be used in other models than the model used for the interpretation of the experimental results.

5.4.3 FIBRE PULL-OUT AND INTERFACIAL MODELLING

Direct experimental evaluation of the nature of the fibre–matrix bond in a fibre-reinforced brittle matrix composite is, with very few exceptions, done by interpreting the so-called fibre pull-out test, Gray (1983), see Fig. 5.8 with a suitable model.

One of the few other approaches are described in Krenchel and Shah (1986) where the fibre material–matrix bond was tested in a torsion test where a plastic film (the film which is later cut into fibres) is placed between the two circular surfaces of two steel cylinders and cemented to these using the matrix material as a glue. The surfaces of the steel cylinders are sand-blasted to ensure a better bond between the matrix and the steel than between the matrix and the fibre material. Torsion load is applied to the steel cylinders, see Fig. 5.9, and the torque causing failure of the fibre material–matrix interface is measured. Unfortunately not much analytical work has yet been done in order to provide an interpretation of these tests. Furthermore, it should be noted that the test only applies to fibre–matrix systems where it is possible to obtain the fibre material in the shape of a film as in the case of fibrillated polypropylene fibres.

Analysis of the fibre pull-out test, on the other hand, has received a lot of attention. The modelling approaches range from description of a fibre embedded in an infinite half space analyzed by numerical, semi-analytical, or purely analytical tools (Muki and Sternberg, 1970; Sternberg and Muki, 1970; Luk and Keer, 1979; Phan-Thien, 1980; Phan-Thien and Goh, 1981; Phan-Thien *et al.*, 1982; Stang, 1985; Steil and Hoysan, 1986; Morrison *et al.*, 1988; Marmonier *et al.*, 1988; Selvadurai and Rajapakse, 1990), to other types of analysis such as the shear lag approach (Outwater and Murphy, 1969; Greszczuk, 1969; Lawrence, 1972; Laws *et al.*, 1973; Takaku and Arridge, 1973; Bartos, 1981; Laws, 1982; Gray, 1984; Gopalaratnam and Shah, 1987; Gao *et al.*, 1988; Palley and Stevans, 1989; Sigl and Evans, 1989; Stang *et al.*, 1990; Hsueh, 1990a, 1990b, 1990c).

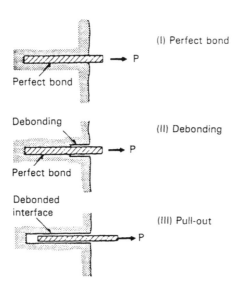

Fig. 5.8 The principle of the single fibre pull-out test. The pull-out process is usually divided into three stages: (I) Perfect fibre–matrix bond, (II) The debonding process, (III) The pull-out process.

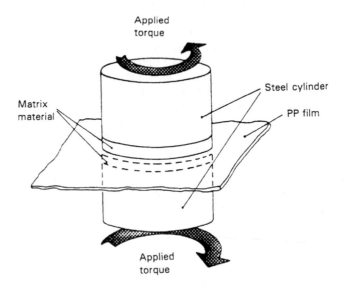

Fig. 5.9 The torsion test used to evaluate the bond between polypropylene fibres and a cementitious matrix. See Krenchel and Shah (1986).

Valuable information regarding the influence of geometrical, physical and interfacial parameters can be gained from simple analytical solutions. Several authors have discussed the applicability of a fracture mechanics criterion for debonding in contrast to a stress-based criterion, see Wells and Beaumont (1982), Wells and Beaumont (1985), Piggot (1987), Shah and Jenq (1987), Stang *et al.* (1990), and Leung and Li (1990).

Takaku and Arridge (1973) derived an approximate solution for the fibre stress needed to extend the debonded zone, also known as the debond stress σ_d, using a stress-based criterion for debonding. Their solution reads:

$$\sigma_d = 2\tau^{crit}\left(\frac{E_f}{2G_m}\right)^{0.5}\left(\ln\left(\frac{b}{r}\right)\right)^2 \tag{5.34}$$

where τ^{crit} is the strength parameter of the interface, E_f is Young's modulus for the fibre, G_m is the shear modulus of the matrix, b is a characteristic length related to the geometry of the fibre and the stiffness of the fibre and the matrix, and r is fibre radius.

Outwater and Murphy (1969) derived a solution for the debond stress using a fracture mechanics approach. They found that

$$\sigma_d = \left(\frac{4E_f G_i^{crit}}{r}\right)^{0.5} \tag{5.35}$$

where G_i^{crit} is the critical energy release rate of the interface.

Stang *et al.* introduced both a stress-based and a fracture mechanics-based criterion for debonding in the same analysis which included frictional stresses τ^* on the debonded interface.

The solution using the stress-based criterion reads:

$$\sigma_d = 2\tau^* \frac{a}{r} + 2\tau^{crit} \left(\frac{E_f \pi}{k_m}\right)^{0.5} \tanh\left(\left(\frac{k_m}{E_f \pi}\right)^{0.5} \frac{(L-a)}{r}\right) \tag{5.36}$$

where L is the embedment length of the fibre, a is the length of the debonded zone, and k_m is the stiffness of the shear lag. Note that the latter solution is similar to the solution by Takaku and Arridge (1973) for vanishing friction and length of the debonded zone.

The solution using the fracture mechanics-based criterion reads:

$$\sigma_d = 2\tau^* \frac{a}{r} + \left(2\tau^* \left(\frac{E_f \pi}{k_m}\right)^{0.5} + \left(\frac{4 E_f G_i^{crit}}{r}\right)^{0.5}\right) \tanh\left(\left(\frac{k_m}{E_f \pi}\right)^{0.5} \frac{(L-a)}{r}\right) \tag{5.37}$$

This latter solution reduces to the solution of Outwater and Murphy (1969) in the case of vanishing friction and large embedment length compared to the fibre radius and the length of the debonded zone.

Clearly the two different debonding criteria predict different parametric dependencies for the debond stress. Especially it is interesting to note, that for vanishing friction on the debonded interface a $1/\sqrt{r}$ size effect should be observed in the strength parameter τ^{crit}. Any deviation from the $1/\sqrt{r}$ size effect can according to equations (5.36) and (5.37) be explained by the presence of frictional stresses on the debonded interface. If the pull-out process is completely dominated by friction no size effect in the strength parameter will be observed. In this case the model predicts that the strength parameter is identical to the frictional stress on the debonded interface.

Wells and Beaumont (1982), Wells and Beaumont (1985), and Shah and Jenq (1987) investigated the fibre radius size effect and found that experimentally observed fibre radius size effect can be explained using the fracture mechanics-based debonding criterion. Piggott (1987) investigated the two criteria by looking at the debond stress–embedment length relationship, and concluded that the solution based on the stress criterion could only describe the experimental results with unrealistically high strength parameters.

5.5 Discussion

The development of models for interfaces in cementitious composites is basically following two trends: first, developments of models which divide the interface into a perfectly bonded and a debonded part, and secondly, development of models which treat the interface as a cohesive interface at all times.

If a model belongs to the perfectly bonded–debonded interface type two types of debonding criteria can be found in the literature: the stress-based criteria and the fracture mechanics-based criteria. The stress-based approach describes the strength of the interfacial bond in terms of one or more strength parameters related to the stress state at the interface while the fracture mechanics approach describes the strength of the bond in terms of toughness parameters.

The more consistent approach seems to be the fracture mechanics approach since the toughness parameters, as opposed to the strength parameters, can be considered to be 'true' material parameters, that is:

- The toughness parameters determined by applying a given model to experimental results do not depend critically on the accuracy and complexity of the analytical model.
- Size effects observed in the strength parameters can be explained by applying a fracture mechanics approach to the debonding problem.

It should also be noted that the fracture mechanics approach to perfectly bonded–debonded interfaces seems to be the prevailing approach used in the modelling of the mechanical behaviour of ceramic matrix composites.

It is at present not clear whether the perfectly bonded–debonded interface approach or the cohesive interface approach is the more appropriate to use in connection with cementitious composite materials. The main difference between the two approaches lies in the fact that the cohesive interface approach assumes that no adhesion exists at the interface while the perfectly bonded–debonded interface approach assumes that stress can be transferred at the interface without any relative displacement taking place at the interface.

It is very difficult to make a direct experimental evaluation of this difference in perception. One attempt to do this is described in Bien (1986) and Bien and Stroeven (1988) where a holographic interferometry technique was used in order to investigate in-plane displacement discontinuities in a pull-out experiment with a steel fibre model embedded in a cementitious matrix. This experiment shows that the interface in this composite system apparently can be divided into a perfectly bonded and a debonded zone.

It is generally accepted that frictional stresses do exist on the debonded interfaces in most cementitious composite systems, thus when the debonded zone is large enough it must be expected that there is not much difference between the perfectly bonded–debonded interface approach and the cohesive interface approach. However, in situations where the debonded zone is very small, the relative displacements at the interface are very small a difference between the predictions of the two approaches must be expected. This situation occurs at the onset of matrix cracking. Thus further insight in the applicability of the two approaches can be gained by applying these types of models in investigations of the first cracking stress/strain of cementitious composites.

5.6 References

Aboudi, J. (1987) Damage in composites - modelling of imperfect bonding, *Composite Science and Technology*, Vol. 28, pp. 103ff.

Abudi, J. (1989) Micromechanical analysis of fibrous composites with coulomb frictional slippage between the phases, *Mechanics of Materials*, Vol. 8, Nos 2 and 3, pp. 103–15.

Aveston, J., Cooper, G.A. and Kelly, A. (1971) Single and multiple fracture, *The Properties of Fibre Composites*, Conference Proceedings, National Physical Laboratory, November, IPC Science and Technology Press Ltd, pp. 15–26.

Atkinson, C., Avila, J., Betz, E. and Smelser, R.E. (1982) The rod pull out problem, theory and experiment, *Journal of the Mechanics and Physics of Solids*, Vol. 30, No.3, pp. 97–120.

Barenblatt, G.I. (1962) The mathematical theory of equilibrium cracks in brittle fracture, *Advances in Applied Mechanics*, Vol. 7, 5129.

Barnes, B.D., Diamond, S. and Dolch, W.L. (1978) The contact zone between Portland cement paste and glass 'aggregate' surfaces, *Cement and Concrete Research*, Vol. 8, pp. 233–244.

Bartos, P. (1980) Analysis of pull–out tests on fibres embedded in brittle matrices, *Journal of Materials Science*, Vol. 15, pp. 3122–8.

Bartos, P. (1981) Review paper: Bond in fibre reinforced cements and concretes, *International Journal of Cement Composites and Lightweight Concrete*, Vol. 3, pp. 159–77.

Beaumont, P.W.R. and Aleszlca (1978) Polymer concrete dispersed with short steel fibres, *Journal of Materials Science*, Vol. 13, pp. 1749–60.

Beneviste, Y. (1984) On the effect of debonding on the overall behaviour of composite materials, *Mechanics of Matererials*, Vol. 3, pp. 349–58.

Beneviste, Y. (1985) The effective mechanical behaviour of composite materials with imperfect contact between the constituents, *Mechanics of Materials*, Vol. 4, pp. 197–208.

Beneviste, Y. and Aboudi, J. (1984) A continuum model for fibre reinforced materials with debonding, *International Journal of Solids and Structures*, Vol. 20, pp. 935–51.

Bentur, A., Diamond, S. and Mindess, S. (1985a) Cracking process in steel fiber reinforced cement paste, *Cement and Concrete Research*, Vol. 15, pp. 331–42.

Bentur, A., Diamond, S. and Mindess, S. (1985b) The microstructure of the steel fibre-cement interface, *Journal of Materials Science*, Vol. 20, pp. 3610–20.

Bentur, A. and Mindess, S. (1990) *Fibre Reinforced Cementitious Composites*, Elsevier Applied Science, (E & FN Spon), UK.

Bien, J. (1986) Holographic Interferometry Study of the Steel Concrete Bond in Pull-Out Testing, Report 1-86-9, Delft University of Technology, Stevin Laboratory.

Bien, J. and Stroeven, P. (1988) Holographic interferometry study of debonding between steel and concrete, *Engineering Applications of Ne70 Composites*, (eds S.A. Paipetis and G.C. Papanicolaou), Omega Scientific, pp. 213–8.

Bowling, J. and Groves, G.W. (1979) The debonding and pull-out of ductile wires from a brittle matrix, *Journal of Materials Science*, Vol. 14, pp. 431–42.

Budiansky, B., Hutchinson, J.W. and Evans, A.G. (1986) Matrix fracture in fiber-reinforced ceramics, *Journal of the Mechanics and Physics of Solids*, Vol. 34, No. 2, pp. 167–89.

Budiansky, B. and Amazigo, J.C. (1989) Toughening by aligned, fictionally constrained fibres, *Journal of the Mechanics and Physics of Solids*, Vol. 37, No. 1, pp. 93–109.

Burakiewicz, A. (1978) testing of fibre bond strength in cement matrix, *Testing and Test Methods of Fibre Cement Composites*, (ed. R.N. Swamy), RILEM Symposium, The Construction Press, UK. pp. 355–65.

Chen, Y.-C. and Hui, C.-Y. (1990) Load transfer in a composite containing a broken fiber with imperfect bonding, *Mechanics of Materials*, Vol. 10, pp. 161–72.

Cook, J. and Gordon, J.E. (1964) A mechanism for the control of crack propagation in all brittle systems, *Proceedings of the Royal Society*, Vol. 282A, pp. 508–20.

de Vekey, R.C. and Majumdar, A.J. (1968) Determining bond strength in fibre reinforced composites, *Magazine of Concrete Research*, Vol. 20, No. 65, pp. 229–34.

Dollar, A. and Stief, P.S. (1988) Load transfer in composites with a coulomb friction interface, *International Journal of Solids and Structures*, Vol. 24, pp. 789–803.

Gao, Y.C. (1987) Debonding along the interface of composites, *Mechanics Research Communication*, Vol. 14, No. 2, pp. 67–72.

Gao, Y., Mai, Y.W. and Cotterell, B. (1988) Fracture of fiber-reinforced materials, *Journal of Applied Mathematics and Physics*, (ZAMP) Vol. 39, pp. 550–72.

Gopalaratnam, V.P. and Shah, S.P. (1987) Tensile failure of steel fiber reinforced mortar, *ASCE Journal of Engineering Mechanics*, Vol. 113, pp. 635–52.

Gray, R.J. (1983) Experimental techniques for measuring fibre/matrix interfacial bond strength, *International Journal of Adhesion and Adhesives*, Vol. 3, pp. 197–202.

Gray, R.J. (1984) Analysis of the effect of embedded fibre length on fibre debonding and pull-out from an elastic matrix. Part 1: Review of theories, *Journal of Materials Science*, Vol. 19, pp. 861–70.

Greszczuk, L.B. (1969) Theoretical studies of the mechanics of the fibre-matrix interface in composites, *Interfaces in Composites*, American Society of Testing and Materials, ASTM STP 452, Philadelphia, pp. 42–58.

Gurney, C. and Hunt (1967) Quasi-static crack propagation, *Proceedings of the Royal Society of London*, Vol. A299, pp. 508–24.

Hamoush, S.A. and Salami, M.R. (1990) Interfacial strain energy release rate of fiber reinforced concrete based on bond stress–slip relationship, *ACI Structural Journal*, Vol. 87, No. 6, pp. 678–86.

Hannant, D.J., Highest, D.C. and Kelly, A. (1983) Toughening of cement and other brittle solids with fibres, *Philosophical Transaction of the Royal Society of London*, Vol. A310, pp. 175–90.

Hillerborg, A. Modeer, M. and Peterson, P.-E. (1976) Analysis of crack formation and crack growth by means of fracture mechanics and finite elements, *Cement and Concrete Research*, Vol. 6, No. 6, pp. 773–81.

Hillerborg, A. (1980) Analysis of fracture by means of the fictitious crack model, particularly for fibre reinforced concrete, *International Journal of Cement Composites and Lightweight Concrete*, Vol. 2, No. 4, pp. 177–84.

Hsueh, C.-H. (1990) Interfacial debonding and fiber pull-out stresses of fiber-reinforced composites, *Materials Science and Engineering*, Vol. A123, No. 1, pp. 1–11.

Hsueh, C.-H. (1990) Interfacial friction analysis for fibre-reinforced composites during fibre push-down (indentation), *Journal of Materials Science*, Vol. 25, No. 2A, pp. 818–28.

Hsueh, C.-H. (1990) Evaluation of interfacial shear strength, residual clamping stress and coefficient of friction for fiber-reinforced ceramic composites, *Acta Metallurgica*, Vol. 38, No. 3, pp. 403–9.

Hutchinson, J.W. (1987) Crack tip shielding by micro-cracking in brittle solids, *Acta Metallurgica*, Vol. 35, No. 7, pp. 1605–19.

Hutchinson, J.W. and Jensen, H.M. (1990) Models of fiber debonding and pullout in brittle composites with friction, *Mechanics of Materials*, Vol. 9, pp. 139–63.

Jensen, H.M., Hutchinson, J.W. and Kim, K.-S. (1990) Decohesion of a cut prestressed film on a substrate, *International Journal of Solids and Structures*, Vol. 26, Nos 9/10, pp. 1099–114.

Jensen, H.M. (1990) Mixed Mode Interface Fracture Criteria, DCAMM Tecnical University of Denmark, Report No. 404.

Korczynskyj, Y., Harris, S.J. and Morley, J.G. (1981) The influence of reinforcing fibres on the growth of cracks in brittle matrix composites, *Journal of Materials Science*, Vol. 16, pp. 1533ff.

Krenchel, H. and Shah, S.P. (1986) Synthetic fibres for tough and durable concrete, *FRC 86. Developments in Fibre Reinforced Cement and Concrete*, RILEM Symposium, (eds R.N. Swamy, R.L. Wagstaffe and D.R. Oakley), Vol. 1., pp. 333–8.

Lawrence, P. (1972) Some theoretical considerations of fibre pull-out from an elastic matrix, *Journal of Materials Science*, Vol. 7, pp. 1–6.

Laws, V., Lawrence, P. and Nurse, R.W. (1973) Reinforcement of brittle matrices by glass fibres, *Journal of Physics D: Applied Physics*, Vol. 6, pp. 523–37.

Laws, V. (1982) Micromechanical aspects of the fibre-cement bond, *Composites*, Vol. 13, pp. 145–51.

Leung, K.Y. and Li, V.C. (1990) Strength-based and fracture-based approaches in the analysis of fiber debonding, *Journal of Materials Science Letters*, Vol. 9, pp. 1140–2.

Lhotellier, F.C. and Brinson, H.F. (1988) Matrix-fiber stress transfer in composite materials: elasto-plastic model with all interphase layer, *Composite Structures*, Vol. 10, pp. 281–301.

Luk, V.K. and Keer, L.M. (1979) Stress analysis for an elastic half space containing an axially-loaded rigid cylindrical rod, *International Journal of Solids and Structures*, Vol. 15, No. 10, pp. 805–27.

Mandel, J.A., Wei, S. and Said, S. (1987) Studies of the properties of the fiber matrix interface in steel fiber reinforced mortar, *ACI Materials Journal*, Vol. 84, No. 2, pp. 101–9.

Marmonier, M.F., Desarmot, G., Barbier, B. and Letalenet, J.M. (1988) A study of the pull-out test by a finite element method (in French), *Journal of Theoretical and Applied Mechanics*, Vol. 7, pp. 741–65.

Marshall, D.B., Cox, B.N. and Evans, A.G. (1985) *Acta Metallurgica*, Vol. 33, pp. 2010ff.

Mori, T. and Mura, T. (1984) An inclusion model for crack arrest in fiber reinforced materials, *Mechanics of Materials*, Vol. 3, pp. 193–8.

Morrison J.K., Shah S.P. and Jena. Y.-S. (1988) Analysis of fiber debonding and pullout in composites, *ASCE Journal of Engineering Mechanics*, Vol. 114, No. 2, pp. 277–94.

Muki, R. and Sternberg, E. (1970) Elastostatic load-transfer to a half-space from a partially embedded axially loaded rod, *International Journal of Solids and Structures*, Vol. 6, No. 1, pp. 69-90.

Naaman A.E. and Shah, S.P. (1976) Pull-out mechanism in steel fiber-reinforced concrete, *Journal of the Structural Division, ASCE*, Vol. 102, pp. 1537–48.

Nammur, G., and Naaman A.E. (1989) Bond stress model for fiber reinforced concrete based on bond stress–slip relationship, *ACI Materials Journal*, Vol. 86, No. 1, pp. 45–57.

Needleman, A. (1987) A continuum model for void nucleation by inclusion debonding, *Journal of Applied Mechanics*, Vol. 54, pp. 525–31.

Needleman, A. (1990a) Analyses of interfacial failure, *Applied Mechanics Reviews*, Vol. 43, No. 5, Part 2, pp. S274–S275.

Needleman, A. (1990b) Analysis of decohesion along an imperfect interface, *International Journal of Fracture*, Vol. 42, No. 1, pp. 21–40.

Needleman, A. (1990c) An analysis of tensile decohesion along an interface, *Journal of the Mechanics and Physics of Solids*, Vol. 38, pp. 289–324.

Needleman, A. and Nutt, S.R. (1989) Void formation in short-fiber composites, *Advances in Fracture Research*, (eds K. Salama *et al.*), Pergamon Press, pp. 2211–20.

Nutt, S.R. and Needleman, A. (1987) Void nucleation at fiber ends in Al-SiC composites, *Scripta Metallurgica*, Vol. 21, pp. 705–10.

Outwater, J.D. and Murphy, M.C. (1967) On the fracture energy of uni-directional laminates, *Proceedings of the the Annual Technical Conference of the Reinforced Plastics/Composites Division*, The Society of the Plastics Industry, Washington, D.C., 11-C-1 – 11-C-8.

Pagano, N.J. and Tandon, G.P. (1990) Modelling of imperfect bonding in fiber reinforced brittle matrix composites, *Mechanics of Materials*, Vol. 9, pp. 49–64.

Page, C.L. (1982) Microstructural features of interfaces in fibre cement composites, *Composites*, Vol. 13, pp. 140ff.

Palley, I. and Stevans, D. (1989) Fracture mechanics approach to the single fiber pull-out problem as applied to the evaluation of the adhesion strength between the fiber and the matrix, *Journal of Adhesion Science and Technology*, Vol. 3, No. 2, pp. 141–53.

Phan-Thien, N. (1980) A contribution to the rigid fibre pull–out problem, *Fibre Science and Technology*, Vol. 13, pp. 179–86.

Phan-Thien, N. and Goh, C.J. (1981) On the fibre pull–out problem, *Journal of Applied Mathematics and Mechanics* (ZAMM), Vol. 61, pp. 89–97.

Phan-Thien, N., Pantelis, G. and Bush, M.B. (1982) On the elastic fibre pullout problem: asymptotic and numerical results, *Journal of Applied Mathematics and Physics*, (ZAMP), Vol. 33, pp. 251–65.

Piggott, M.R., Chua, P.S. and Andison (1985) *Polymer Composites*, Vol. 6, pp. 242–8.

Piggott, M.R. (1987) Debonding and friction at fibre-polymer interfaces. I: Criteria for failure and sliding, *Composite Science and Technology*, Vol. 30, pp. 295–306.

Pinchin, D.J. and Tabor, D. (1978a) Interfacial phenomena in steel fibre reinforced cement I. Structure and strength of the interfacial region, *Cement and Concrete Research*, Vol. 8, pp. 15–24.

Pinchin, D.J. and Tabor, D. (1978b) Inelastic behaviour in steel wire pull-out from portland cement mortar, *Journal of Materials Science*, Vol. 13, pp. 1261–6.

Pinchin, D.J. and Tabor, D. (1978c) Interfacial contact pressure and frictional stress transfer in steel fibre cement, *Testing and Test Methods of Fibre Cement Composites*, RILEM Symposium, (ed. R.N. Swamy), The Construction Press, UK, pp. 337–44.

Proctor, B.A. (1990) A review of the theory of GRC, *Cement and Concrete Composites*, Vol. 12, pp. 53–61.

Selvadurai, A.P.S. (1983) Concentrated body force loading of an elastically bridged penny shaped flaw in a unidirectional fibre reinforced composite, *International Journal of Fracture*, Vol. 21, pp. 149–59.

Selvadurai, A.P.S. and Rajapakse, R.K.N.D. (1990) Axial stiffness of anchoring rods embedded in elastic media, *Canadian Journal of Civil Engineering*, Vol. 17, pp. 321–8.

Shah, S.P. and Jenq, Y.-S. (1987) Fracture mechanics of interfaces, *Bonding in Cementitious Composites*, (eds S. Mindess and S.P. Shah), Materials Research Society, Vol. 114.

Sigl, L.S. and Evans A.G. (1989) Effects of residual stress and frictional sliding on cracking and pull-out in brittle matrix composites, *Mechanics of Materials*, Vol. 8, pp. 1–12.

Sih, G.C. and Rice, J.R. (1964) The bending of plates of dissimilar materials with cracks, *Transactions,*

ASME, Series E, Journal of Applied Mechanics., Vol. 31, pp. 477–82.

Stang, H. (1985) The Fibre Pull-Out Problem: An Analytical Investigation, Series R, No. 204. Department of Structural Engineering, Technical University of Denmark.

Stang, H. and Shah, S.P. (1986) Failure of fibre-reinforced composites by pullout fracture, *Journal of Materials Science, Vol.* 21, No. 3, pp. 953–7.

Stang, H. (1987) A double inclusion model for microcrack arrest in fibre reinforced brittle materials, *Journal of the Mechanics and Physics of Solids,* Vol. 35, No. 3, pp. 325–42.

Stang, H., Li, Z. and Shah, S.P. (1990) The pullout problem. Stress versus fracture mechanical approach, *Journal of Engineering Mechanics, ASCE,* Vol. 116, No. 10, pp. 2136–50.

Steif, P.S. and Hoysan, S.F. (1986) On load transfer between imperfectly bonded constituents, *Mechanics of Materials,* Vol. 5, pp. 375–82.

Stern, M. and Hong, C.C. (1976) Stress intensity at a crack between bonded dissimilar materials, *Advances in Engineering Science,* Proceedings, 13th Annual Meeting, Society of Engineering Science, Nampton, 1976, NASA CP-2001, Vol. 2. pp. 699–710.

Sternberg, E. and Muki, R. (1970) Load-absorption by a filament in a fiber-reinforced material, *Journal of Applied Mathematics and Physics,* (ZAMP), Vol. 21, pp. 553–69.

Takaku, A. and Arridge, R.G.C. (1973) The effect of interfacial radial and shear stress on fibre pull-out in composite materials, *Journal of Physics D: Applied Physics,* Vol. 6, pp. 2038–47.

Tvergaard, V. (1989) Effect of Fibre Debonding in a Whisker-reinforced Metal, DCAMM Technical University of Denmark, Report No. 400.

Wang, Y., Li, V.C. and Backer, S. (1988) Modelling of fibre pull-out from a cement matrix, *International Journal of Cement Composites and Lightweight Concrete,* Vol. 10, No. 3, pp. 143–9.

Wang, Y., Backer, S., and Li, V.C. (1989) A statistical tensile model of fibre reinforced cementitious composites, *Composites,* Vol. 20, No. 3, pp. 265–74.

Wei, S., Mandel, J.A. and Said, S. (1986) Study of the interface strength in steel fibre reinforced cement-based composites, *Journal of the American Concrete Institute, Proceedings,* Vol. 83, No. 4, pp. 597605.

Wells, J.K. and Beaumont, P.W.R. (1982) Fracture energy maps for fibre composites, *Journal of Materials Science,* Vol. 17, pp. 397–405.

Wells, J.K. and Beaumont, P.W.R. (1985) Debonding and pull-out processes in fibrous composites, *Journal of Materials Science,* Vol. 20, pp. 1275–84.

PART THREE

INFLUENCE OF TRANSITION ZONE
ON BEHAVIOUR OF CONCRETE

6

Influence of the interfacial transition zone on composite mechanical properties

J.C. Maso

6.1 Introduction

This chapter covers relationships between the interfacial transition zone microstructure and the mechanical properties of composites. The latter are as follows:

- the reversibility limit and internal mechanisms governing the limit being crossed,
- breaking,
- the elastic modulus,
- Poisson's ratio,
- viscous and ageing components in behaviour resulting from application of long duration mechanical actions (creep and relaxation),
- crack propagation mechanisms.

The composites concerned are 'conventional' concretes or lightweight aggregate concretes, high performance concretes and reinforced concretes (using steel reinforcement or fibres). 'Conventional' concretes, which represent the majority of cement matrices, either with or without steel reinforcement, are made up of non-porous aggregates that are less subject to deformation than the cement slurry coating them at all stages of hardening, and mechanically more resistant. These aggregates are in most cases non-reactive with the cement. In some cases, however, they may have either a positive or negative action whose effects can be positive or negative (alkali reaction). For these materials, except for cases of alkali reaction and for the most commonly used cements (CPA and CPJ), we can consider the main interfacial transition zone characteristics that will determine the composite's response to mechanical actions to be known. Lightweight aggregate concretes are made using natural or artificial aggregates with large open porosity, more subject to deformation and mechanically less resistant than the cement matrix coating them after sufficient hardening time. High performance concretes are distinguishable from 'conventional' concretes by the incorporation of water-reducing additives and sub-micronic solid additives (silica fumes to date).

For lightweight aggregate concretes and high performance concretes, microstructural studies of the interfacial zone are rarer than those relating to 'conventional' concretes with non-reactive aggregates. However, we have enough data available to explain the differences observed in mechanical behaviour.

For the reinforcement–concrete association, whether this involves oxidised, non-oxidised or galvanised steel, we have sufficient knowledge of the interfacial zone microstructure for

Interfacial Transition Zone in Concrete. Edited by J.C. Maso. RILEM Report 11.
Published in 1996 by E & FN Spon, 2–6 Boundary Row, London SE1 8HN. ISBN 0 419 20010 X.

CPA and CPJ cement. However, we have practically no experimental basis to enable us to relate the reinforced composite's mechanical response to the reinforcement–concrete interface's microstructure. In what follows, we shall ignore fibre materials and reinforced concrete.

6.2 Reversibility limit and irreversibility mechanisms

6.2.1 CONVENTIONAL CONCRETES WITH CHEMICALLY NON-REACTIVE AGGREGATES

The interfacial transition zone's main characteristics in this case involve greater porosity in the vicinity of aggregates at all stages of hardening and the presence of larger and better formed crystals than in the bulk of the slurry. These are portlandite crystals that are readily cleavable.

Let us consider the reversibility limit surface trace on one of the load area reference planes p_1, p_2, p_3 (Fig. 6.1) and analyse the micro-mechanisms that will lead to this surface being crossed on an elementary model made of an inclusion and featuring an aggregate, in a hydrated cement matrix (Fig. 6.2).

Under compression single-axis stress, along direction 2, to simplify matters, the mechanical properties of continuous media shows a tightening of the slurry's isostatic components against the aggregate, due to differences in tendency to deform (the aggregate is less subject to deformation than the hydrated cement slurry). On the samples centred on A and A', the interfacial transition zone is compressed, while it is stretched in the central part between

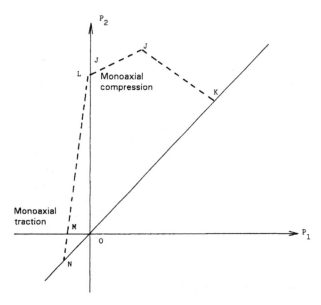

Fig. 6.1 Trace of the surface of limit of reversibility in a plane (p_1, p_2, 0) of the space of loading for a normal concrete.

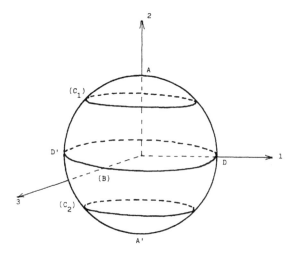

Fig. 6.2 Elementary model representing an aggregate in cementitious matrix and the directions of loading 1, 2 and 3.

circles (C_1) and (C_2). Compression is at its maximum on A and A' while tensile stress is at its maximum along the diametrical circle (B). Shearing on the interface, which is null on A and A' and along the circle (B) is at its maximum along circles (C_1) and (C_2).

The first irreversibilities in single-axis compression are due to collapsing of the interfacial transition zone in the compression samples. Micro-fractures result therefrom, with a reduction in porosity. This is logical as the interfacial transition zone is more porous than the rest of the matrix and because it is subjected to a higher level of stress in the areas around points A and A'.

If we add compression along direction 1, it will reduce the compressive stresses on the samples surrounding points A and A', as also tensile stress due to p_2 on the samples centred on D and D', but will increase the tensile stress due to p_2 over the rest of the interface. In these conditions, it is logical for us to have to increase the intensity of compression p_2 as the intensity of compression increases in direction 1 to remain at the limit of reversibility (segment IJ, Fig. 6.1).

The phenomenon involving a crossover to irreversibility by consolidation in the most compressed areas of the interfacial transition zone is followed by crossing the limit by micro-cracking. Indeed, as the compression intensities p_1 and p_2 increase, the tensile stresses on the interface in direction 3 and the shearing stresses will also increase. There will then be a crossover to irreversibility through cracking along the sides of the aggregate (segment JK, Fig. 6.1) and p_2 will have to diminish as p_1 increases up to total compression.

If, on the other hand, the load in direction 1 is tensile stress, crossover to irreversibility by interfacial transition zone micro-cracking (segment IL, Fig. 6.1) will very quickly succeed the phenomenon involving crossover to irreversibility through consolidation of the interfacial transition zone in its most compressed parts. Indeed, when p_1 is tensile stress, two tensile loads on the interface will be added in the samples around points D and D': that due to compression in direction 2 and that due to tensile stress in direction 1. The same applies to shearing, which reaches maximum values on the (p_1, p_2 plane) where the compressed and stretched parts meet. On the macroscopic level, we shall then be able to observe a very rapid decrease in the intensity of p_2 at the limit of reversibility as p_1 is seen to increase (segment LM, Fig. 6.1).

For $p_2 = 0$ (point M), stress is by single-axis tensile stress in direction 1. Irreversibility is due to micro-cracking of the interfacial transition zone. When tensile stress is applied in direction 2, it tends to reduce tensile stresses due to p_1 in the samples around points D and D'. As a logical consequence, it can then be observed in Fig. 1 that the value for p_1 for which irreversibility will be reached will increase with p_2 (segment MN, Fig. 6.1).

As concrete is isotropic - unless a special casting technique is used - the trace for the reversibility limit will be symmetrical in relation to the first bisector. For the same reason, traces will be the same on the two other load space reference planes. There is a ternary symmetry for the reversibility limit surface around the trisector.

Under the assumption that we can consider the mechanical behaviour of the aggregates and the slurry to be elastic-linear, which is reasonable considering the low load level at reversibility limit, we shall be able to determine mathematically the trace for the experimental reversibility limit surface for Fig. 6.1. This will correspond to the intersection of two traces: the first corresponding to the mechanism involving collapsing by compression of the interfacial transition zone and the second corresponding to the mechanism involving cracking of the interfacial transition zone by tensile stress and shearing (Fig. 6.3) (Maso, 1969; M'Saadi, 1981; Maso, Romeu and M'Saadi, 1984).

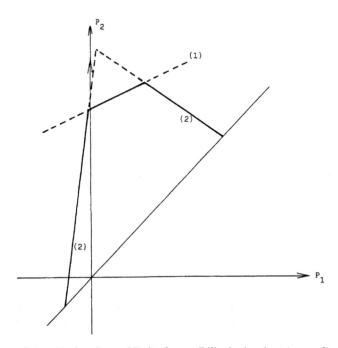

Fig. 6.3 Traces of theoretical surfaces of limit of reversibility in the plane $(p_1, p_2, 0)$ of the space of loading for the elementary model: (1) trace of the limit of reversibility by crushing of the interfacial transition zone; (2) trace of the limit of reversibility by microcracking of the interfacial transition zone.

By calculation, using an analogous model (Fig. 6.4), we can also find the bending reversibility mean. This will be reached for the composite when the inclusion interfacial transition zone cracking occurs in the stretched zone. The mathematical model and the experimental model results coincide. Cracking of the interfacial transition zone can be explained using either a tensile stress Rankine formula local criterion, a shearing local criterion or a Mohr-Cacquot local criterion associating tensile stress and shearing (Rossi, Acker and Boulay, 1985).

The interfacial transition zone collapsing phenomenon has been confirmed using physical models which show an increase in the interfacial transition zone's abrasiveness with compression intensity (Fig. 6.5) (Pons, Maso and Chouicha, 1988).

6.2.2 LIGHTWEIGHT AGGREGATE CONCRETES

The phenomenon involving suction of water from the slurry when the aggregates are initially dry or diffusion phenomena when the aggregates are initially saturated, combined with the considerable irregularity of surface areas due to aggregate pore openings on the surface (significantly reducing the local curve radii) completely modify the interfacial transition zone microstructure. We shall no longer observe either higher porosity than in the rest of the slurry or crystals with greater dimensions as in the case of non-porous aggregates. We shall therefore no longer find a zone that is more sensitive to the generation and propagation of cracks in the immediate vicinity of the surface, indeed, the opposite.

In addition, and this constitutes a major difference, the aggregates are more liable to deformation and less resistant than cement slurry, due to their porosity.

As these materials are considerably more fragile than conventional concretes, the reversibility limit and breaking surfaces on the load space reference planes practically merge (Fig. 6.6).

For the reasons indicated above (except of course for mechanical stress close to compressive collapse), irreversibility will here always be reached by micro-cracking. Here too, the phenomena involved can be translated using the same basic model as above.

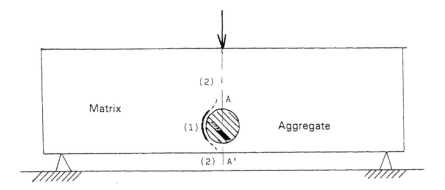

Fig. 6.4 Limit of reversibility for a three bending point model by microcracking of the interfacial transition zone. By increasing the loading, the cracking appears first at the matrix–aggregate interface and remains at this interface until the load become high enough to propagate into the matrix (2).

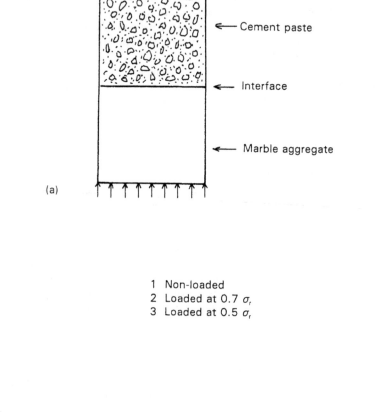

(a)

(b)

Fig. 6.5 Abrasivity of the interfacial transition zone as a function of loading: (a) Experimental model; (b) Curves of abrasivity as a function of the distance to the interface.

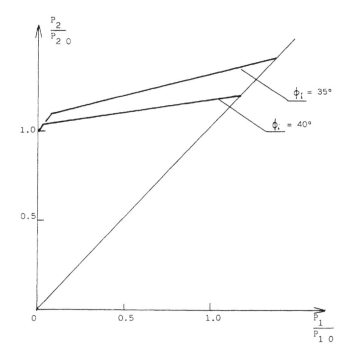

Fig. 6.6 Traces of the surfaces of limit of reversibility in biaxial compression for concrete of lightweight aggregates. $p_{10} = p_{20}$: limit of reversibility in biaxial compression. ϕ_i angle of internal friction. Coulomb–Mohr Criteria.

Theoretical traces are introduced on the same figure as for experimental traces. The breaking criterion will be a Mohr-Caqot local criterion (M'Saadi, 1981; Maso, Romeu and M'Saadi, 1984).

6.2.3 HIGH PERFORMANCE CONCRETES

High performance concretes are obtained by densifying the matrix. This densification is particularly marked in the immediate vicinity of the associated materials' surfaces. This densification is obtained by adding water-reducing additives either alone or associated with sub-micronic grain solid additives. Porosity gradients will then be increasingly reduced as near-complete filling of existing spaces between aggregates and cement grains is achieved. Further, as the space is divided, we shall no longer have large dimension crystals or preferential orientations.

To our knowledge, there are no studies on such concrete's response to multi-axial stress. However, we can surmise that irreversibilities by consolidation do not exist and that crossing the reversibility line will only occur through micro-cracking (except, of course, for stresses close to triaxial collapse). We shall find an indirect confirmation in the analysis of stress–deformation curves.

6.3 Behaviour beyond the reversibility breaking limit

6.3.1 CONVENTIONAL CONCRETES

The wall effect generating water/cement ratio gradients and, as a result, maximum porosity depends on the respective radii of aggregates and cement grains in contact with them. It will therefore be more pronounced where the aggregates are large and less pronounced when the aggregates are small (sand). This has not been demonstrated experimentally but we can assume it to be the case. Irreversibilities by crushing of the interfacial transition zones would be seen to occur in the vicinity of the largest aggregates then in the vicinity of the smaller aggregates, in order of decreasing size. Similarly, irreversibilities by cracking would be seen to occur first at the interfaces with the largest aggregates and would then gradually affect smaller sized grains before bridging through the part of the matrix made up of cement slurry and smaller sized aggregates.

When tensile stress is predominant, we quickly reach localisation of cracking leading to breaking. When compression is predominant, crack propagation is much more gradual, with the final stage corresponding to breaking of the matrix bridges separating interfacial cracks.

In the present case, the breaking surfaces are logically orientated perpendicular to the main tensile stress. In the second case, they will also logically be orientated parallel to the direction of compression. In both cases, the breaking surfaces will follow the contour of the largest aggregates.

Gradual deterioration of concretes beyond the reversibility limit thus corresponds and more gradually with compression (Maso, Romeu and M'Saadi, 1984).

The first non-linear characteristics for compression stress–deformation curves are almost certainly due to gradual crushing of the interfacial transition zones. This crushing will lead to an additional increase in longitudinal deformation in relation to the purely linear condition (Fig. 6.7). Micro-cracking followed by cracking will then follow as a cause of non-linearities.

6.3.2 LIGHTWEIGHT AGGREGATE CONCRETES

We have stated that irreversibility always arises by cracking. We are not aware of work that tells us whether the first micro-cracks appear in the aggregates or in the matrix.

In addition, breaking surfaces no longer follow the contours of aggregates, whatever the nature of the stress (at least after adequate hardening time). We should here consider that the cement slurry plays its adhesive role, with breaks no longer occurring on the interfaces with the associated materials (Maso, Romeu and M'Saadi, 1984).

The near linear character up to breaking in a load build-up or up to a peak in deformation is certainly due to the fact that there is no longer an appearance then sequential propagation of cracking and these concretes are logically more fragile when subjected to the same type of stress as conventional concretes.

6.3.3 HIGH PERFORMANCE CONCRETES

As the interfacial zone is densified, compressive consolidations will be delayed then suppressed. The stress–deformation curve should then linearise in its central part, in prolongation of its initial part. This is indeed what can be observed. Lateral cracking of the

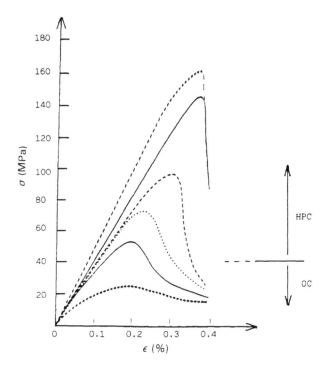

Fig. 6.7 Strain–stress curves for ordinary concretes (OC) and high performance concrete (HPC).

interfacial zone subsequent to compressive consolidation is also delayed and we should here also see linearisation of the upper part of the stress deformation curve. This linearisation is clearer and of greater extent the greater the densification.

While for conventional concretes, the fracture surfaces always follow the aggregate contours, the break surfaces for high performance concretes may follow the aggregate contours: this will be the case for concretes made using rocks with very high mechanical resistance (diabase, for example). They may, on the other hand, cross the aggregates (this is the case for limestones).

From an adequate densification level of the interfacial zone, the concrete will behave in practically the same way as a diphasic polygranular medium with two mechanically homogeneous phases and we should be able to model it as such.

6.4 Elastic modulus and Poisson's ratio

6.4.1 Conventional concretes

Several models have been proposed for a diphasic material's design: the series model, the parallel model and mixed models. This is no doubt due to the fact that an elastic modulus equal to that of a slurry with the same water/cement ratio which has hardened far from any wall to the cement slurry.

Current computation facilities should allow the real microstructure to be linked with the elasticity modulus, thanks to the porosity gradients in the concrete due to interfaces.

In compression, the modulus tangent to the stress–deformation curve in monotonic load will decrease immediately after the first irreversibilities due to crushing of interfacial transition zones; indeed, the latter lead to an increase in longitudinal deformations. On the other hand, if you unload after the reversibility limit has been reached, it can be observed that the modulus tangent to unloading together with the tangent for original re-loading are greater than the initial tangent modulus at the origin of the first load (Pons and Maso, 1984).

The appearance of microcracking will accelerate the modulus' diminution tangent to the monotonic load curve. If you carry out an unload–reload cycle after the micro-cracking threshold, the moduli tangent to the unload origin and at the reload origin will be lower than the tangent modulus at the first load's origin (Fig. 6.8b).

This clearly shows that irreversibilities due to crushing of interfacial transition zones, by reducing porosity here corresponds to a consolidation as there is rigidification, while micro-cracking corresponds to damage to the material.

However, Poisson's ratio will remain practically constant throughout the consolidation phase whereas it increases very rapidly as soon as micro-cracking appears (Fig. 6.9).

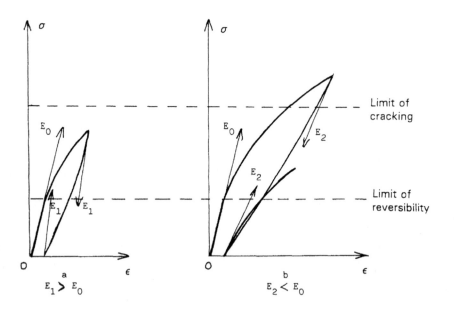

Fig. 6.8 Evaluation of the modulus of elasticity as a function of loading for ordinary concrete, in monoaxial compression: (a) the level of loading is higher than the limit of reversibility but lower than the limit of cracking; (b) the level of loading is higher than the limit of cracking.

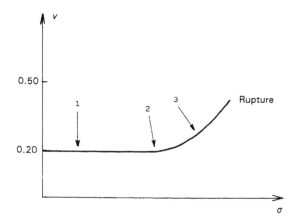

Fig. 6.9 Poisson's ratio in monoaxial compression: (1) phase of crushing of interfacial transition zones; (2) beginning of microcracking; (3) phase of propagation of cracks through the matrix.

6.4.2 LIGHTWEIGHT AGGREGATE CONCRETES

Here, we do not observe an increase in the modulus over cycles before cracking. The only phenomenon to appear will be damage in the vicinity of the stress–deformation curve peak. This damage will accelerate very quickly, thus explaining the fragile nature of this type of concrete. This concurs with what we know about the microstructure of these concretes, particularly at the interfaces.

Poisson's ratio shows a similar change to that for conventional concretes. This is logical, as this parameter is not influenced by consolidation of interfacial transition zones.

6.4.3 HIGH PERFORMANCE CONCRETES

In compression, the modulus will change less with loading insofar as the densification of interfacial transition zones is greater. The more significant densification is, the more it will increase the elastic-linear behaviour field area. The modulus' change will only appear a little before the stress–deformation curve peak. This signifies damage, which will be more rapid insofar as densification is significant, bearing witness to the gradual fragilisation of the concrete with densification of interfacial transition zones and disappearance of larger scale crystals.

We do not have sufficient data available here to connect the microstructure at the interfaces with the macroscopic response to stationary loads in an analysis by type of concrete, as we were able to do previously for the other mechanical structures. The tight coupling between creep and retraction makes analysis particularly delicate.

For a load over 28 days with a chronology specific to the material, the creep of high performance concrete should be less than for conventional concretes, and this will be more pronounced as resistance increases. The creep coefficient, of the order of 2 for conventional concretes in unlimited time, would be reduced to 1.5 or even 1 for concretes with characteristic resistance after 28 days equal to 60 MPa and would be less than 1 for concretes with characteristic resistance after 28 days equal to 100 MPa (Fig. 6.10).

We could conclude from the above analyses that densification of the interfacial transition zone leads to a reduction in creep. However, some authors, where they obtain a lower creep value for high performance concretes for a load over 28 days and at 40% of the breaking load, obtain a result opposite by 75% to the breaking load. Further, for loads over 1 day, high performance concrete creep will always be greater due to the higher creep speed at the start of loading. This would also remain as true for a relative humidity of 50% as for absence of exchange with the outer environment and in a saturated environment.

Here, clearly, there is a new field for research in terms of the property–microstructure relation.

6.5 Crack propagation mechanisms

For conventional concretes, the most commonly accepted assumption used to reconcile the physical impossibility of infinite stress in the vicinity of the crack bottom and the special nature of the breaking stress field in linear mechanics corresponds to the process zone. The process zone itself seems to be made up of a network of micro-cracks in the ligament which propagate with the crack bottom.

Recent experiments demonstrate the existence of discontinuous micro-cracking at the interfaces on the crack trajectory. As the crack spreads in a three point bending experiment, the stresses increase in the ligament, leading to cracking of the interfacial transition zones (Fig. 6.10) (Bascoul and Turatsinze, 1992).

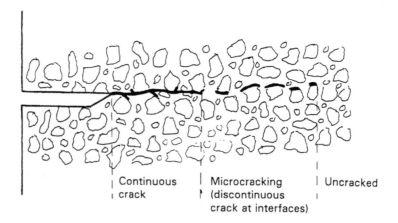

Continuous	Microcracking	Uncracked
crack	(discontinuous	
	crack at interfaces)	

Fig. 6.10 Propagation of crack in mode I of opening in a three point bending test. Process zone.

The process zone thus clearly exists, but is limited to a discontinuous cracking on the crack trajectory. The rapid variation in intensity of stresses when you move away from the maximum moment section explains the fact that as soon as an interfacial transition zone is cracked it can no longer develop another micro-crack on the same scale in the ligament. Further, due to the tortuous nature of the cracked zone, the crack's opening cannot occur without friction. The corresponding friction forces play a similar role to that played by Barenblatt cohesion forces. We can thus observe that crack propagation in mode I in conventional concrete is directly governed by the existence and the specific microstructure of interfacial transition zones.

6.6 Aggregate reactivity

We should never lose sight of the fact that in concretes, hydration reactions occur in confined spaces with small dimensions. When they come into contact with cement solutions, most rock minerals generally used in cement making are superficially dissolved and, over several tens of angstroms, the ion content from these minerals can be significant enough to lead to hydrated compounds by combination with ions from anhydrous cement compounds. The most commonly known example is where there is superficial dissolution of limestone aggregates leading, with CPA cements, to the formation of intermediate solid solutions (Farran, 1956). While the existence of such layers that are highly adhesive to the aggregate have been observed, they still need to be clearly identified (Javelas, Maso, Ollivier and Thenoz, 1975). In addition, this superficial dissolution of aggregates limits, with CPAs, the formation of portlandite crystals, which is a favourable element from the mechanical point of view.

Here, clearly, there is an explanation of the fact that micro-cracks do not usually develop on the surface itself of aggregates, but further away.

This type of reactivity can be said to be positive with respect to mechanical properties and it might even be advantageous to encourage it.

There is a second type of reactivity which can be said to be negative. This is the case of alkali-reaction and interference between the interfacial transition zone microstructure and the silica-alkali reaction which certainly needs to be dealt with in greater depth.

6.7 Conclusion

The interfacial transition zone plays an essential role in the properties of conventional concretes: elastic modulus, reversibility limit, micro-crack generation and crack propagation process, together with mechanical resistance.

All of which, on a level with the material formulation, tends to reduce, indeed to suppress the interfacial transition zone characteristics: maximum porosity, bigger and better formed crystals, preferential orientations, logically tends to increase the performance of each of these mechanical properties. However, it has a cost: the increase of the brittleness (high performance concrete and very high performance concrete).

6.8 References

Bascoul, A. and Turatsinze, A. (1992) Discontinuous crack growth as fracture process zone through SEM Analysis, *Fracture Mechanics of Concrete Structures*, Proceedings 1st International Conference, FraMCOS 1, Breckenridge, USA, Elsevier Applied Science, London, 1992.

Farran, J. (1956) Contribution minéralogique à l'étude de l'adhérence entre les constituants hydrates des ciments et les matériaux enrobés, *Matériaux et Constructions*, Nos 490-491, pp. 155–72, No. 492, pp. 191–209.

Javelas, R., Maso, J.C., Ollivier, J.P. and Thenoz, B. (1975) Observation directe au microscope électronique par transmission de la liaison pâte de ciment-granulats dans des mortiers de calcite et de quartz, *Cement and Concrete Research*, Vol. 5, pp. 285–94.

Maso, J.C. (1969) La nature mineralogique des aggrégats, facteur essentiel de la resistance des bétons à la rupture et à l'action du gel, *Matériaux et Constructions*, No. 647-648-649.

Maso, J.C., Romeu, G. and M'Saadi, R. (1984) Modelisation de la réponse des bétons à des sollicitations biaxiales jusqu'au seuil de fissuration, Colloque International RILEM-CEB-CNRS *Réponse du Béton aux Sollicitations Multiaxiales*, Vol. II, Toulouse 22-24 May.

M'Saadi, R. (1981) Etude sur modèles de la limite de réversibilité des matériaux granulaires cohérents: cas ou l'irréversibilité est due à la rupture de la liaison entre la matrice et les inclusions. These Toulouse.

Pons, G., Maso, J.C. and Chouicha, J. (1988) Application d'une méthode d'usure par abrasion à la détermination du comportement mécanique de l'auréole de transition d'un béton soumis à des chargements cycliques, RILEM, *Matériaux et Constructions*, Vol. 21, No. 123, May.

Pons, G. and Maso, J.C. (1984) Microstructure evolution of concrete under low-frequency cyclic loading: determination of the porosity variations, *Advances in Fracture Research*, ICF6, New Delhi, December.

Rossi, P., Acker, P. and Boulay. C. (1985) Observation et Identification de la Microfissuration des Bétons Hydrauliques. Rapport AFREM, October.

7

The effect of the transition zone on transfer properties of concrete

J.P. Ollivier and M. Massat

7.1 Introduction

The durability of a concrete structure is largely dependent on the possibilities of ingress of water, gases and ions into the porous material. Consequently, the microstructural characteristics of the transition zone affect the transport properties of concrete. The most important parameters controlling penetration of external agents are:

- the pores and the microcracks size and their connectivity;
- the nature of the phases in this zone and their reactivity with the chemical species.

In this chapter, the influence of the transition zone on the permeability and diffusion are discussed. Firstly, a theoretical approach is presented in order to link the microstructural parameters of concrete into the transition zone and the transport coefficients.

7.2 Transport properties and microstructure
(Ollivier and Massat, 1991)

7.2.1 PERMEABILITY AND MICROSTRUCTURE

The permeability of a porous medium is a global coefficient which characterizes the flow of a fluid caused by a pressure head. This coefficient in an intrinsic parameter of the material. Assuming laminar flow, the permeability is given by Hagen–Poiseuille law according to the equation:

$$Q = \frac{K}{\mu} \times A \times \frac{dP}{dZ} \tag{7.1}$$

Q	= volume of fluid flowing in unit time;
K	= permeability of the medium;
μ	= viscosity of the fluid
dP/dZ	= pressure gradient

Interfacial Transition Zone in Concrete. Edited by J.C. Maso. RILEM Report 11.
Published in 1996 by E & FN Spon, 2-6 Boundary Row, London SE1 8HN. ISBN 0 419 20010 X.

The permeability of concrete is function of its pore structure which depends on composition, curing history and the damage from physical, mechanical or chemical aggressions. The flow properties also depend on the moisture content of the pore system (Nilsson, 1991); this last aspect of the problem is not discussed here. As previously mentioned by Scheidegger (1974), it is obvious that no simple correlation between porosity and permeability exists. The correlation that might be looked for is between *structure* and *permeability*.

The problem of establishing correlations between a measurable parameter such as permeability and a general concept such as structure is dealt with by modelling the porous structure of the solid and the flow into the material.

7.2.1.1 The Carman–Kozeny theory
The porous space of the medium is assumed to be equivalent to a straight channel of constant cross-section a, (Fig. 7.1). The Carman–Kozeny equation is established on the basis of three other assumptions.

1. The Dupuit–Forchheimer assumption, relating the pore velocity v_e, the apparent velocity v_a of the fluid and the porosity p:

$$v_a = pv_e = \frac{a}{A} \times v_e \tag{7.2}$$

where A is the cross-section of the material.

2. The velocity of the fluid in the equivalent channel under the pressure gradient $\Delta P/L_e$ is given by:

$$v_e = \frac{m^2}{h_0 \cdot \mu} \cdot \frac{\Delta P}{L_e} \tag{7.3}$$

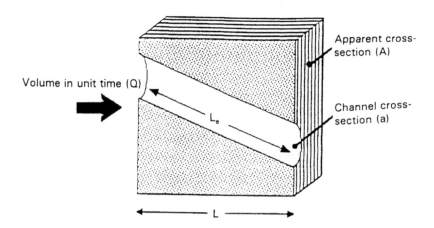

Fig. 7.1 Schematic representation of the equivalent medium model (from David, 1991).

where μ is the viscosity of the fluid, h_0 a form factor of the channel and m its hydraulic radius. This hydraulic radius is defined as the ratio between the volume of the channel (the same as the volume of the porous space) and the surface of the channel (aL_e : the same as the surface of the pores). h_0 is a geometric term which depends on the shape of the cross-section.

3. The apparent velocity may be related to the permeability (K) according to Darcy's law:

$$v_a = \frac{K}{\mu}.\frac{\Delta P}{L} \qquad (7.4)$$

where L is the thickness of the material in the flow direction.

By introducing the specific surface area S_v (the area of the pores by apparent volume unit of the material) and the tortuosity of the porous space, $T = (L_e/L)^2$, the resulting Carman–Kozeny formula is:

$$K = \frac{pm^2}{h_0 T} = \frac{1}{h_0 T}.\frac{p^3}{S_v(1 - p^2)} \qquad (7.5)$$

If the h_0 term fluctuates very little with the shape of the channel, the tortuosity T varies widely with the topography of the porous space and the predicting power of this approach is limited. Nevertheless, the Carman–Kozeny equation is popular because the pre-factor $h_0 T$ is roughly constant for powders ($h_0 T = 5$); it is easy to calculate the specific surface area from permeability measurements (Blaine or Lea and Nurse methods).

For concrete or cement paste the Carman–Kozeny equation is not valid because the very delicate microstructure of hydrates and the resulting high specific surface area cannot be easily correlated with the equivalent channel dimension relevant for flow (Garboczi, 1990). Another reason of the collapse of this theory is the non-respect of one of its implicit assumptions: the pores are not uniform in size in either undamaged or microcracked materials. Attempts to link permeability of hardened cement paste or concrete and hydraulic radius have to be mentioned but the results tend to disagree with the Carman–Kozeny formulation:

For high performance concrete, a relationship is given by:

$$\log K = 38.45 + 4.08 \log \tau \ m^2 \qquad (7.6)$$

where τ is the specific porosity (Nyame and Illston, 1981).

For concrete

$$K = T.p^\beta.m^a \qquad (7.7)$$

This last formula is not useful for prediction because T and β depend on the cement type (Watson and Oyeka, 1981).

7.2.1.2 Statistical modelling
Statistical modelling involves quantitative data on the size and spacial arrangement of pores

or cracks. Unfortunately, the geometry of the porous space in cement paste and concrete is extremely complex. Calculations of permeability thus involve modelling pore and crack geometry: usually the shape of the pores is represented by cylindrical pipes (capillaries) and thin circular discs representing cracks; a laminar regime is assumed for the fluid flow. Models described in the literature relate to different hypotheses by their spacial arrangement of pipes or discs.

(a) Capillary models
Pores are represented by several parallel capillaries. Different variants have been described by Scheidegger (1974): permeability may be calculated as a function of porosity and pore size (diameter) distribution. For example, in the straight capillary model in which the porous medium is represented by a bundle of straight, parallel capillaries of uniform diameter δ, the permeability is given by:

$$K = \frac{p\,\delta^2}{32} \tag{7.8}$$

δ may be considered as the 'average' pore diameter of the porous medium.

This modelling presents two major disadvantages: flow is anisotropic and the connectivity of the pores is not taken into account. This approach has been modified by considering an isotropic distribution of pipes but improved results are obtained by statistical analysis.

(b) Statistical models (Gueguen and Diennes, 1989)
By modelling pores by means of a set of capillaries (Fig. 7.2) of variable radii r and isotropically distributed lengths λ, the permeability is statistically calculated as:

$$K = \frac{\pi}{32} \cdot n_0\, \lambda\, r^4 \tag{7.9}$$

where n_0 is the number of capillaries per unit volume and r^4 the fourth-order moment of the radius distribution.

In a simplified version, an approximate expression of r^4 is used. The radius distribution is

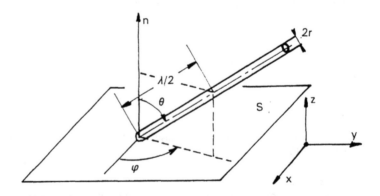

Fig. 7.2 Capillaries model: notations.

assumed to be narrow and an average spacing between capillaries, l, is introduced so that $n_0 = 1/l^3$. Considering that only a fraction f, of connected capillaries participate in the volume flow, Gueguen and Dienes (1989) have calculated the following expression:

$$K = \frac{\pi}{32} . f\lambda . \frac{r^4}{l^3}$$ (7.10)

The porosity of the medium may be linked to the previous microscopic data according to

$$p = \frac{\pi r^2 \lambda}{l^3}$$ (7.11)

The following expression is thus obtained:

$$K = \frac{fr^2 p}{32}$$ (7.12)

This is very similar to equation (7.8) and to the Carman–Kozeny relationship in which the equivalent channel would be an r-radius cylindrical pipe. In that condition the tortuosity T, is related to the fraction of connected capillaries: $f = 4/T$. Nevertheless, this modelling is more efficient than the previous one because the term f is physically defined and percolation theory enables it to be calculated as a function of the pipe arrangement.

A similar approach has been carried out for fractured media (Fig. 7.3). Dienes (1982) has shown that an isotropic distribution of cracks with radius c, number density n_0, and aspect ratio $A = w/c$, results in a permeability:

$$K = \frac{4\pi}{15} . A^3 n_0 c^5 f$$ (7.13)

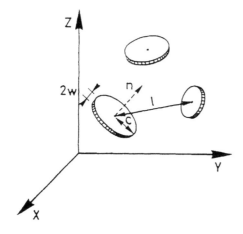

Fig. 7.3 Crack model: notations.

The f factor accounts for the fraction of connected cracks which participate in the volume flow.

By assuming narrow distributions of the structural parameters and introducing the porosity of the material due to cracks, p, the permeability is found to be:

$$K = \frac{4\pi}{15} \cdot \frac{fw^3c^2}{l^3} = \frac{2}{15} \cdot fw^2p \qquad (7.14)$$

At this stage of the modelling, it is possible to analyse the influence of the microscopic parameters $(r, l, \lambda, w$ and $c)$ on the permeability of the medium. For calculation of permeability, the evaluation of the term f which describes the connection of pipes or cracks is necessary. According to the percolation theory (Stauffer, 1985), f depends on the probability P, that two pipes or two cracks intersect, which is itself a function of the microscopic characteristics and of the spacial arrangement of pipes or cracks.

The great complexity of this arrangement in concrete leads to another step of modelling: the centres of pipes or cracks are located on the nodes (sites) of a geometrical network. Below a 'percolation threshold' P_c, flow across the media is impossible $(f = 0)$ and for $P > P_c$ permeability increases rapidly $(f \rightarrow 1)$. The percolation threshold depends on the geometrical characteristics of the network. For example, with a Bethe network where each site has 4 neighbours, $P_c = 1/3$ (Fig. 7.4).

Moreover, probability P, can be calculated as a function of the microscopical characteristics of capillaries or cracks. Consequently, the percolation threshold condition leads to another relationship:

- for capillaries dispersed on the previous network:

$$\lambda^2 \cdot \frac{r}{2} \cdot l^3 > \frac{1}{3} \qquad (7.15)$$

- for cracks on the same network:

$$\frac{\pi^2 c^3}{4 l^3} > \frac{1}{3} \qquad (7.16)$$

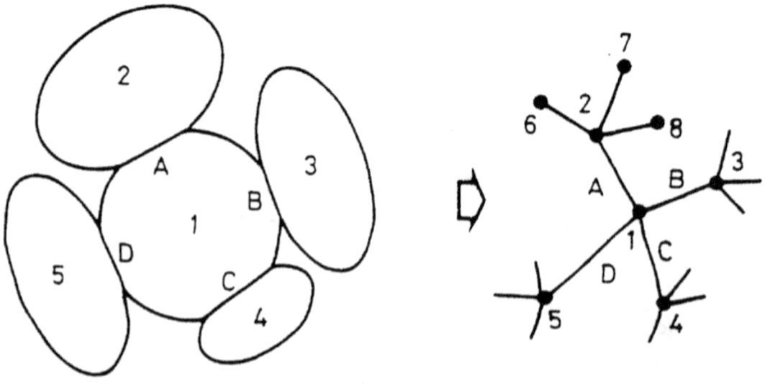

Fig. 7.4 Analysis of connected cracks by a Bethe network (from Gueguen and Diennes, 1989).

For a fractured medium, this last relation can be presented in morphological terms: the flow across the material is possible if the average diameter is greater than the average distance separating the cracks.

The fraction of interconnected capillaries or cracks increases rapidly for $P > P_c$ and the percolation theory gives:

- the proportion P of sites belonging to a continuous path between two opposite sides of the network (infinite cluster): $P \propto (P - P_c)^\beta$
- the term f (in equations 7.2 and 7.3) in the vicinity of P_c: $f \propto (P - P_c)^\mu$
- the permeability of the medium: $K \propto (P - P_c)^\mu$

$\propto$ stands for proportional and the critical exponents μ and β depend on the dimensions of the network but are independent of its geometry: for example, in three dimensions, $\mu = 2$ and $\beta = 0.4$ (Fig. 7.5).

This model has been successfully tested on rocks (David, 1991; Gueguen, David and Darot, 1986) assuming $f = 1$. If all the capillaries or cracks are not connected it is necessary to complete this model to calculate the permeability. The formation factor F is then defined by:

$$F = \frac{\sigma}{\sigma_0} = \frac{D}{D_0} \tag{7.17}$$

where σ_0 and D_0 are the fluid conductivity and diffusion coefficient; σ and D are the conductivity and diffusion coefficient of the medium.

A calculation of F gives $Ffp = 4$; measurement of the formation factor can determine f and calculate the permeability from the equation:

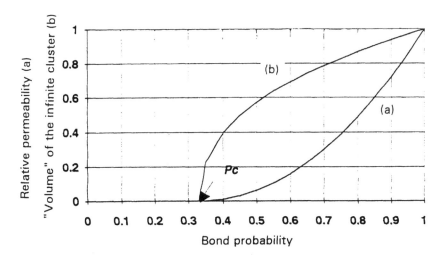

Fig. 7.5 Normalised 'volume' of the infinite cluster $P(P)/P_0$ and its permeability K/K_0 (from Charlaix, 1987).

$$K = \frac{r^2}{8F} \qquad \text{for capillary model} \qquad\qquad (7.18)$$

$$K = \frac{8w^2}{15F} \qquad \text{for crack model} \qquad\qquad (7.19)$$

This model is unable accurately to predict concrete permeability because the porous characteristics are quite different from those of the model in the sense that pores and cracks may exist simultaneously. Moreover, the pore size distribution is very wide and preferential paths for fluid flow may exist in the strongly connected porosity.

7.2.1.3 Katz–Thompson model

Based on percolation theory, the Katz–Thompson model (Katz and Thomson, 1986) establishes a relationship between the permeability K, the formation factor F, and the size of a critical path d_c, of a porous specimen:

$$K = \frac{d_c^2}{226F} \qquad\qquad (7.20)$$

In the Katz–Thompson approach, fluid flow and the conduction of ions across the porous specimen are determined by the same dimension d_c called the critical pore diameter: d_c is the dimension of the first continuous pathway created by a non-wetting fluid invading a porous specimen initially saturated by a wetting fluid. This dimension can be related to the breakthrough pressure P_b (hydrostatic pressure necessary for creating this continuous pathway across the specimen) according to the Laplace equation:

$$P_b = 4A \cdot \frac{\cos\theta}{d_c} \qquad\qquad (7.21)$$

where A is the surface tension between the two immiscible fluids and θ the contact angle.

d_c is traditionally determined by using the mercury porosimetry curve of the specimen: the breakthrough is assumed to correspond to the inflexion point of the curve (Fig. 7.6).

Good agreement between Katz–Thompson theory and measured values have been found for rocks (Gueguen, David and Darot, 1986) and by Garboczi (1990) for hardened cement paste. Measuring d_c was difficult in specimens which were too large for commercially available equipment. Further criticism of this experimental approach is that the mercury intrusion process needs initial drying of the specimen; cracking and breakthrough may be affected (Feldman, 1986). With mortars (sand/cement ratio = 3) the d_c value is more than one order of magnitude greater than for cement paste: Fig. 7.7. According to the Katz–Thomson model the mortar/cement paste permeability ratio should be greater than 100. In fact this ratio is found (US Bureau of Reclamation, 1975) to be only 2 for the same water/cement ratio (0.6).

For cracked media the theory still holds, but for prediction of cracked concrete permeability, it may rendered more effective in determining breakthrough (and d_c) by using a water-expulsion method with equipment similar to the Bremont porosimeter (Bremont, 1957). With such a technique, the specimen is water-saturated and an increasing gas pressure

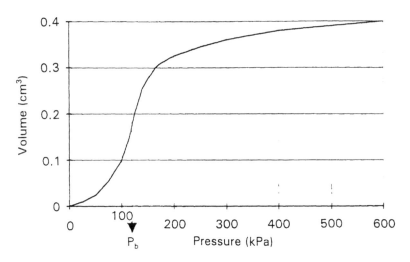

Fig. 7.6 Determination of the breakthrough point from a mercury porosimetry experiment.

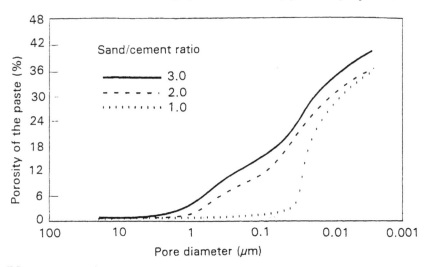

Fig. 7.7 Mercury porosimetry of mortar and cement paste (Katz and Thomson, 1986).

is applied on one side. The breakthrough pressure is measured when the first gas bubble appears on the opposite side.

7.2.2 DIFFUSION AND MICROSTRUCTURE

Transport by diffusion of a chemical species is due to concentration gradients. The transport of ions in a specified liquid is characterized by the diffusion coefficient D_f which is defined by Fick's first law:

$$J_x = -D_f \cdot \frac{dC}{dx} \tag{7.22}$$

where J_x is the flow of the ionic speciaes in the x direction and dx its concentration gradient in the same direction. By considering Fick's first law, it appears that the diffusion kinetic is independent of the dimension of the channels in which it takes place.

In a porous material, transport by diffusion takes place through the connected porosity and it depends greatly of the nature and the quantity of the filling fluid. In a porous medium, Fick's second law may be used to describe the change of concentration with time:

$$\frac{\partial C}{\partial t} = \frac{\partial}{\partial x} \left(D_e \cdot \frac{\partial C}{\partial x} \right) \tag{7.23}$$

D_e is the effective diffusion coefficient of the ion species in the material. It takes into account the characteristics of the pores in terms of three parameters: the porosity p, the tortuosity T, and the constrictivity τ. This last parameter gives an account of the cross-section variations of the pores. D_e is given by Van Brakel and Heertjes (1974):

$$D_e = D_f \cdot \frac{\tau p}{T} \tag{7.24}$$

In this relation the dimension of the channels does not appear clearly. An important factor seems to be the geometrical arrangement of the pores.

When an interaction occurs (chemical reaction, adsorption, etc) between the chemical species and the porous medium, D_e is simply replaced by D_a, the apparent diffusion coefficient, which is a function of time. D_e is generally called the 'diffusivity' of the material; its value is usually measured in diffusion cells and it is the one which is reported in papers concerning diffusion characteristics.

7.3 Transition zone and permeability

7.3.1 INFLUENCE UPON THE PERMEABILITY OF CONCRETE

Theoretically, adding low permeability aggregates to cement paste should reduce overall permeability by interrupting canal continuity in the cement paste matrix, particularly in the case of young paste with a high water/cement ratio, and high capillary porosity. In consequence, concrete and mortar of the same age and water content should be less permeable than pure paste. Results of some tests, however, indicate that the opposite is true. Data in Fig. 7.8 show the considerable increase in permeability when aggregates are added to a paste or mortar (Katz and Thomson, 1986). The greater the size of the aggregates, the greater is the permeability. Watson and Oyeka (1981), by measuring oil permeability, observe that permeability of concrete specimens is about 100 times greater than for cement paste specimens.

These results have been analysed by Mehta (1986) who attributes the permeability increase to the presence of microcracks in the transitional zone at the interface between paste and

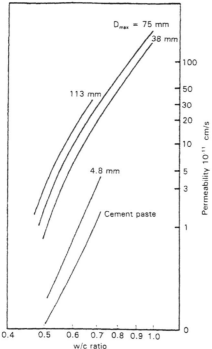

Fig. 7.8 Influence of water/cement ratio and maximum aggregate size on concrete permeability (Katz and Thomson, 1986).

aggregates, the interface cracking becoming more severe as the maximum aggregate size increases. The effect of microcracks in the interfacial zone on permeation is extremely important because the permeability varies with the third exponent of their opening.

On the contrary, Dhir *et al.* (1989) found no significant difference in air permeabilities of concrete made from aggregates of differing sizes (up to 20 mm). The slight permeability increase observed when using 40 mm aggregates is considered to be due to the lower quality of the paste–aggregate interface.

In fact the analysis of the influence of aggregates characteristics such as grading or content on the permeability of concrete is not easy because different composition parameters are generally varying together in concrete. For example in Garboczi (1990) the design strength, water/cement ratio and slump are constant as the maximum aggregate size varies but, evidently, the cement content decreases as D_{max} increases. By supposing a more permeable area around the coarser aggregates, this effect may be masked by a lower volume of paste in the concrete.

The analysis of aggregate effect on permeability may be completed by Nyame's results (1985). The permeability of normal and lightweight mortars is varying with the aggregate volume concentration and the porosity as shown on Figs 7.9 and 7.10. Nyame concludes as follows:

- At increasing aggregate volume concentrations, interfacial effects increase the permeability, whilst absorption of paste water by the aggregate reduces the permeability of mortars.

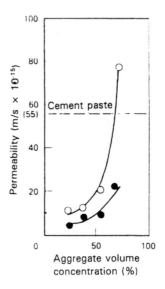

Fig. 7.9 Effect of aggregate volume concentration upon the permeability of mortars (Nyame, 1985).

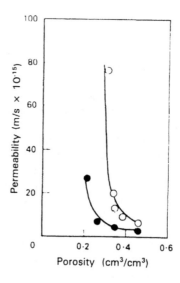

Fig. 7.10 Relation between porosity and permeability of mortars (Nyame, 1985).

- Lightweight mortar is about twice as permeable as, and not of a different order of permeability from, sand mortar at a given aggregate volume concentration.
- The permeability of mortars increases as porosity reduces, contrary to the response of hardened cement pastes.

This last conclusion is not in agreement with the results of Gueguen and Diennes (1989) in which the permeability of concrete increases with the total porosity. As shown in Fig. 7.11 the permeability of concrete increases with increasing aggregate content. The influence of the transition zone may be underlined by considering the data which correspond to the same water/cement ratio, and the same grading (black points on Fig. 7.9): the permeability is really affected by the aggregate content *per se*.

Due to the difficulties in separating the influence of composition parameters on the interface or on the bulk properties, alternative approaches have been attempted to appreciate directly the contribution of the transition zone to the permeability. The results are presented in the next part of this report.

7.3.2 PERMEABILITY OF THE INTERFACIAL ZONE

The water coefficient of permeability of a cement–rock composite has been measured by Skalny and Mindess (1984). The coefficient of permeability was essentially constant with time in tests on quartz and limestone themselves. However, tests on either pure cement paste or on cement–rock composites showed that the 'apparent' water permeability decreases for several days before reaching a stable value corresponding to an established regime in the specimen (Fig. 7.12).

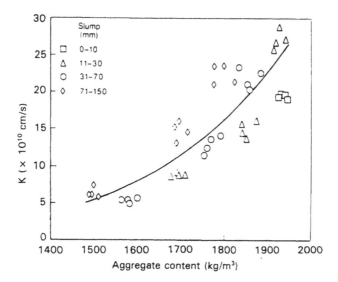

Fig. 7.11 Variation of permeability with aggregate content for OPC concrete (Watson and Oyeka, 1981).

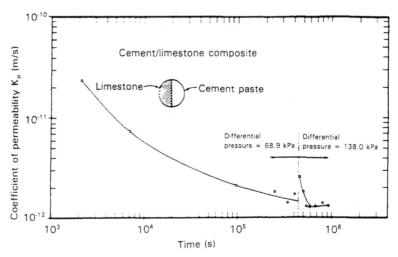

Fig. 7.12 Permeability of cement–limestone composite specimen as a function of time (Skalny and Mindess, 1984).

The test results suggest to the authors that the interfacial region itself does not appear to play a significant role with regard to water permeability but no comparative values of the permeability coefficients are given by Skalny and Mindess (1984). A similar conclusion has been obtained by Wakeley and Roy (1982): the contact zone between different rocks and grouts or mortars does not provide a preferred pathway for water flow.

In an older paper, Valenta (1961) discussing the significance of the bond for the durability of concrete, presented an original device to analyse the water permeability of the interfacial zone. The water transfer is easier as the roughness of the joint increases and is greatly favoured by air curing instead of water curing. This last effect may be explained by cracking at the interface. Unfortunately, no comparative results between the bulk and the interface properties are furnished in this paper.

7.4 Transition zone and ion diffusion

The effect of water/cement ratio of cement paste on chloride diffusivity was found to be relatively small (Goto and Roy, 1981). This is in good agreement with theory which stipulate that diffusivity is independent on the pore size.

The effects of transition zone features on ion diffusivity are difficult to establish because no direct data exist. As compared with hardened cement paste, the order of magnitude of the effective diffusion coefficient of a given species in concrete is about the same: in the range of 10^{-11}–10^{-12} m^2/s with ordinary OPC concrete.

7.5 References

Bremont, P. (1957) Morphology, size and distribution of pores in ceramic products, *Bulletin de Societe Francaise Ceramique,* Vol. 37, pp. 2338.

Charlaix, E. (1987) These de Doctorat de l'Universite Paris VI.

David, C. (1991) La permeabilite et la conductivite electrique des roches dans la croute: experiences en laboratoire et modeles theoriques. These de l'universite L. Pasteur, Strasbourg.

Dhir, R.K., Hewlett, P.C. and Chan, Y.N. (1989) Near surface characteristics of concrete: intrinsic permeability, *Magazine of Concrete Research,* Vol. 41, No. 147, pp. 87–97.

Dienes, J.K. (1982) Permeability, percolation and statistical crack mechanics, *Issues in Rock Mechanics,* (eds R.E. Goodman and F.E. Heuze), American Institute of Mining, Metallurgical and Petroleum Engineers, New York.

Feldman, R.F. (1986) Influence of condensed silica fume and sand/cement ratio on pore structure and frost resistance of Portland cement mortars, *Fly Ash, Silica Fume, Slag and Natural Pozzolans in Concrete,* American Concrete Institute Publication SP-91, pp. 973–89.

Garboczi, E.J. (1990) Permeability, diffusivity and microstructural parameters: a critical review, *Cement and Concrete Research,* Vol. 20, No. 4, pp. 591-601.

Goto, S. and Roy, D.M. (1981) Diffusion of ions through hardened cement pastes, *Cement and Concrete Research,* Vol. 11, pp. 751-7.

Gueguen, Y., David, C. and Darot, M. (1986) Model and time constant for permeability evolution in cracked rocks, *Geophysical Research Letters,* Vol. 13, pp. 460-3.

Gueguen, Y. and Diennes, J. (1989) Transport properties of rocks from statistics and percolation, *Mathematical Geology,* Vol. 21, pp. 460-3.

Katz, A.J. and Thomson, A.H. (1986) a quantitative prediction of permeability in porous rocks, *Physical Reviews B,* Vol. 24, pp. 8179-81.

Mehta, P.K. (1986) *Concrete: Structure, Properties and Materials,* Prentice Hall, Englewood Cliffs, NJ.

Nilsson, L.A. (1995) Relationships between flow coefficients for moisture, water, gases and ions in concrete, *Performance Criteria for Concrete Durability,* (eds J. Kropp and H.K. Hilsdorf), RILEM Report, E & FN Spon, London.

Nyame, B.K. (1985) Permeability of normal and lightweight mortars, *Magazine of Concrete Research,* Vol. 37, No. 130, pp. 44–8.

Nyame, B.K. and Illston, J.M. (1981) Relationships between permeability and pore structure of hardened cement paste, *Magazine of Concrete Research,* Vol. 33, No. 116, pp. 139–46

Ollivier, J.P. and Massat, M. (1991) Permeability and microstructure of concrete: a review of modelling, *XV International Symposium on the Scientific Basis for Nuclear Waste Management,* Strasbourg.

Scheidegger, A.E. (1974) *The Physics of Flow Through Porous Media.* 3rd edition, University of Toronto Press.

Skalny, J.P. and Mindess, S. (1984) Physico-chemical phenomena at the cement paste–aggregate interface, *10th International Symposium on Reactivity of Solids,* Dijon.

Stauffer, D. (1985) *Introduction to Percolation Theory,* Taylor and Francis.

Valenta, O. (1961) The significance of the aggregate–cement bond for the durability of concrete, *International RILEM Symposium Durability of Concrete,* Praha, Preliminary report, pp. 53–87.

Van Brakel, J. and Heertjes, P.M. (1974) Analysis of diffusion in macroporous media in terms of a porosity, a tortuosity and a constrictivity factor, *International Journal of Heat and Mass Transfer,* Vol. 17, pp. 1093-1103.

Wakeley, L.D. and Roy, D.M. (1982) A method for testing the permeability between grout and rock, *Cement and Concrete Research,* Vol. 12, No. 4, pp. 533-4.

Watson, A.J. and Oyeka, C.C. (1981) Oil permeability of hardened cement pastes and concrete, *Magazine of Concrete Research,* Vol. 33, No. 115, pp. 85-95.

U.S. Bureau of Reclamation (1975) *Concrete Manual,* 8th edition, p. 37.

8

Action of environmental conditions

F. Massazza

8.1 Introduction

The constituents of concrete, namely cement paste, aggregate as well as reinforcement, are materials that differ considerably in chemical composition, physical structure and response to external stresses. For this reason at the interface between the different materials the chemical and physical properties show abrupt changes. The interfaces have been the topic of numerous studies and the theme of important scientific meetings:

- The Structure of Concrete and its Behaviour under Load, London, 1965, Cement and Concrete Association;
- International RILEM Colloquium, Liaisons de Contact dans les Materiaux Composites Utilises en Genie Civil, Toulouse, 1972;
- 7th International Congress on the Chemistry of Cement, Paris, 1980;
- International Conference on Bond in Concrete, Paisley, 1982
- International RILEM Colloquium, Liaisons Pâtes de Ciment/Matériaux Associés, Toulouse, 1982;
- 8th International Congress on the Chemistry of Cement, Rio de Janeiro, 1986;

However, direct studies concerning their influence on durability of concrete are rather few. As a matter of fact, the shortage of direct test data is less unfavourable than might appear inasmuch as the factors which influence the durability of concrete as a whole act similarly on the durability of the interface.

A concrete is so much more durable, i.e. it practically retains for so much longer its initial project characteristics,

(i) the better it is able to withstand the penetration of aggressive agents,
(ii) the less it is affected by external ambient conditions, namely temperature, freezing-thawing, humidity.

These requirements clearly apply to all parts of the concrete, and therefore also to the interfaces. Thus, the conclusions drawn by evaluating the factors which affect the durability of concrete may be applied, at least qualitatively, to the interface too.

Interfacial Transition Zone in Concrete. Edited by J.C. Maso. RILEM Report 11.
Published in 1996 by E & FN Spon, 2–6 Boundary Row, London SE1 8HN. ISBN 0 419 20010 X.

8.2 Relation between porosity and permeability

Experience shows that the more durable a concrete is, the more compact and the less permeable to fluids it is. A porous material may be impermeable, but a permeable material must necessarily be porous. This rule also applies to concrete and cement paste which are permeable and hence porous.

There is a strong connection between porosity and permeability (Powers, Copeland, Hayes and Mann, 1954), but the permeability of pastes becomes practically zero when the porosity drops below 35% in volume with regard to both Portland (Powers, Copeland, Hayes and Mann, 1954) and pozzolana cements (Costa and Massazza, 1988). Thus permeability cancels out when porosity values are still far from negligible. This behaviour may be accounted for by admitting that whenever the porosity drops below a certain figure, the system of intercommunicating capillary pores is transformed into a system of closed, or very thin, pores which are thus practically impervious to water (Powers, Copeland and Mann, 1959).

The porosity of cement paste depends on a number of factors, including the type of cement, but the chief ones are the water/cement ratio and curing.

To obtain a workable paste, the water/cement ratio must be at least 0.32, i.e. it should exceed the value required for complete hydration (0.28). By using superplasticizing admixtures, good workability is also achievable with a ratio of 0.26, but in any case, even when the water/cement ratio is equal to 0.28 or less, a part of the water does not combine and it is to be found in a relatively free state in the capillary pores and gel pores. Needless to say, porosity increases as the water/cement ratio increases (Powers, Copeland, Hayes and Mann, 1954).

Suitable curing, i.e. sufficiently prolonged and carried out under conditions of high relative humidity, minimizes both porosity and permeability (see Fig. 8.1) (Costa and Massazza, 1988).

The length of curing depends on the cement type, meaning that rapid hardening cements require shorter curing than slow hardening ones, but the differences are not essential if wet curing lasts longer than seven days and the concrete is compact enough.

Besides depending on the water/cement ratio and curing, porosity and permeability of concrete depend also on the cement content. It is clear that the cement paste, possibly diluted with the finest sand parts, must fill the existing voids as well as envelop every single grain in the aggregate without solution of continuity.

The aggregate has a number of different important and favourable functions in concrete inasmuch as it reduces the consequences of the heat of hydration of cement as well as shrinkage due to drying and creep. Nevertheless, it has an unfavourable feature: it requires water. As a result, to ensure similar workability, concretes have a higher water/cement ratio than pastes. Part of the excess water, if the aggregates are only slightly absorbent, which is usually the case, is found in the paste but part of it is distributed on the paste–aggregate interface where it lubricates the particles of the aggregate. As a consequence, the interfacial zone has a greater porosity and a greater permeability than the bulk cement paste. There is quite a lot of experimental evidence in this respect.

The higher porosity existing in the interfacial zone has been determined on composite concrete specimens consisting of a single cylindrical piece of limestone or quartz surrounded by a ring of cement paste (w/c = 0.4) (Tognon and Cangiano, 1980).

The role played by the interface has been estimated by subtracting the porosity of the two components, aggregate and paste, from the total porosity of the composite specimen. The results quoted in Table 8.1 show that the measured porosity of composite is considerably

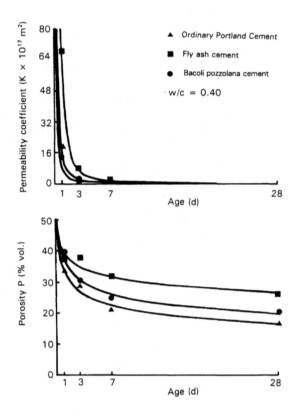

Fig. 8.1 Variation of paste permeability and porosity with time (Costa and Massazza, 1988). OPC = ordinary Portland cement: FAC = fly ash cement; BPC = Bacoli pozzolana cement.

Table 8.1 Porosity of annular concrete composites (Tognon and Cangiano, 1980)

Aggregate	Pore volume (cm³/g)		Difference
	Measured	Calculated	
Limestone	0.0346	0.0252	0.0094
Quartz	0.0463	0.0360	0.0103

higher than calculated porosity and the difference must be ascribed to the paste layer around the aggregate particles.

The greater porosity of the interface is confirmed by the pore size distribution of the annular concrete samples. The curve in Fig. 8.2 shows a peak between 150 and 300 nm which, since it is not observed in the paste or the aggregate, may be reasonably ascribed to the interfacial zone (Tognon and Cangiano, 1980).

The greater porosity of the interfacial zone is associated with a greater permeability. This fact has been proved by determining the moment when a conductive solution containing 3%

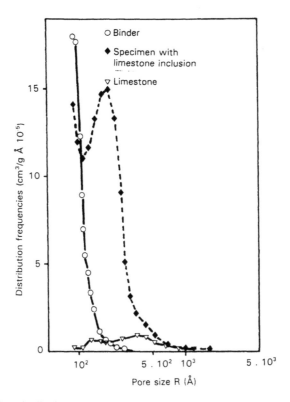

Fig. 8.2 Pore size distribution of limestone inclusion and water-cured binder paste and concrete (Tognon and Cangiano, 1980). Binder = 62.5% rapid hardening cement + 37.5% ground silica.

of NaCl reached the electrodes situated at the interface and in the bulk paste, respectively. In all cases, the electric signal indicating the arrival of the saline solution reached the electrode placed at the interface sooner than the electrode inserted in the bulk cement paste.

Electric conductivity measurements have shown that the interfacial zone is less compact and more permeable than the paste when the aggregate is inert (quartz): the opposite is true when the aggregate is reactive (Ptl cement clinker, aluminous clinker) (Xie Ping and Ming-shu, 1988).

The greater permeability of the interfacial zone compared with the paste was confirmed by comparing the permeability coefficient of neat paste and limestone discs with the coefficient of discs formed by a cylindrical inclusion of limestone surrounded by paste. Data shown in Tables 8.2 and 8.3 (Costa, Facoetti and Massazza, 1990) raise no doubts in this respect. Prior to permeability testing, specimens were dried by removing water from the pores by isopropanol and n-pentane displacement (Marsh, Day, Bonner and Illston, 1983). This method, which is milder than oven- or under-vacuum-drying, prevents cracking in the paste. Among other obvious features, the data shown in Tables 8.2 and 8.3 (Costa, Facoetti and Massazza, 1990) highlight that permeability diminishes, the longer the curing and the lower the water/cement ratio.

Table 8.2 Coefficient of specific permeability ($K = m^2 \times 10^{17}$) measured on samples based on Portland cement (Costa, Facoetti and Massazza, 1990).

Material		Plain paste disks				Disks with limestone core			
Curing (days)		1	3	7	28	1	3	7	28
Without admix.	R w/c 0.32	0.15	0.07	0.04	0.01	3.48	1.15	0.41	0.13
+ 1% NSF		0.31	0.08	0.05	0.01	0.23	0.08	0.06	0.38
Without admix.	R w/c 0.40	1.72	0.33	0.13	0.07	117.50	33.40	10.29	2.53
+ 1% NSF		1.97	0.25	0.10	0.06	5.28	0.78	0.21	0.12

Table 8.3 Coefficient of specific permeability ($K = m^2 \times 10^{17}$) measured on samples based on pozzolanic cement (Costa, Facoetti and Massazza, 1990).

Material		Plain paste disks				Disks with limestone core			
Curing (days)		1	3	7	28	1	3	7	28
Without admixt.	R w/c 0.32	2.08	0.37	0.11	0	3.24	0.51	0.22	0.21
+ 1% NSF		1.63	0.26	0.12	0	2.20	0.21	0.08	0.07
Without admix.	R w/c 0.40	10.11	1.10	0.39	0	15.70	12.82	3.60	0.58
+ 1% NSF		11.15	1.64	0.47	0	14.60	1.58	0.41	0.01

The tables also show that:

1. The permeability of pozzolanic cement pastes, initially higher, is lower after 28 days of curing than that of the control Portland cement pastes.
2. The early permeability of the pastes containing superfluidizer is slightly higher than that of pastes in which the admixture is missing.
3. The final permeability of composites having the same water/cement ratio and containing superfluidizers is instead lower by approximately one order of magnitude than that of composites without admixture.

The modification of permeability caused by the superplasticizer means that the admixture, used in this case not as a water reducer but as a agent dispersing plasticizer, has reduced the thickness of the liquid film covering the aggregate particles of the fresh concrete. This favourable effect on the interface must obviously have matching effects on the durability of the concrete.

8.3 Relation between porosity and strength

It is known that there is a close relationship between strength and porosity in cement pastes (see Fig. 8.3) (Brunauer, 1965).

When cement powders are highly compacted by using low water/cement ratios and heavy pressures, the linear relation between the logarithm of the porosity and the compressive strength has been extended to a value of approximately 700 N/mm^2 (Roy and Gouda, 1973) (see Fig. 8.4).

A similar relationship between compressive strength and porosity is observed in concrete. The rule whereby concrete's compressive strength diminishes by 5–6% for every 1% increase of the porosity has clearly an empirical basis, but confirms the close relationship existing between the two parameters (Wright, 1953).

The compressive strength of ordinary concrete is lower than that of its components since the bond between aggregate and cement paste is weaker than the cohesion between the two constituents, and this bond is weaker because the porosity is greater in the interface than in the paste (see point 8.2).

The relative weakness of the interfacial bond between paste and aggregate has been stressed through direct mechanical tests (flexural, tensile, shear, etc.) performed on aggregate specimens cut and covered with cement pastes (Alexander, Wadlaw and Gilbert, 1968; Shah and Slate, 1968; Hsu and Slate, 1963), microhardness tests (Lyubimova and Pinus, 1962) as well as measurements of the stress intensifying factor. In all these tests the above parameters

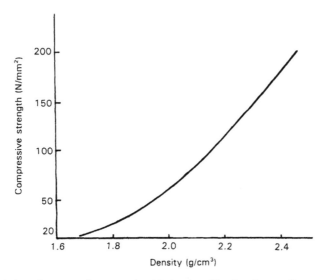

Fig. 8.3 Variation of compressive strength with density of hardened paste (Brunauer, 1965).

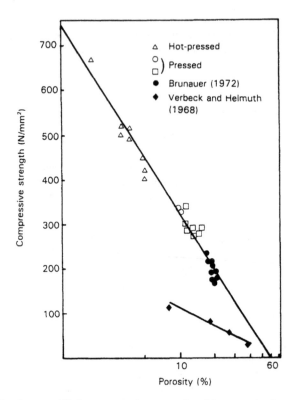

Fig. 8.4 Relation between 28-day compressive strength and log porosity for cement pastes (Roy and Gouda, 1973).

are always definitely lower than those of the paste and the aggregate (Diamond and Gomez-Toledo, 1978).

By improving the strength of the paste, e.g. by reducing the water/cement ratio, not only the concrete's strength improves but also that of the interfacial bond.

Fig. 8.5 (Burg, 1982) shows that a reduction of the total concrete porosity can be obtained by adding increasing quantities of microsilica and by using up to 4% on cement weight of superplasticizer. In this case the action of the microsilica cannot be distinguished from that of the superplasticizer. It is to be noted, however, that although the water/cement ratio remained unchanged, the bond strength has increased noticeably (Bürge, 1982).

The bond stress, measured on mortars according to the ANSI/ASTM standard C234 (ASTM, 1980) method, becomes as much as twofold when a 12–13% substitution of cement with microsilica is effected. As the bond stress doubles, a 50% decrease in porosity occurs. Larger substitutions (>13%) affect the bond stress negatively (Bürge, 1982).

The improvement of the concrete's strength always influences the paste–aggregate and paste–reinforcement bond positively.

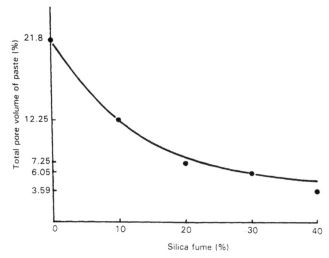

Fig. 8.5 Total pore volume of cement pastes with increasing content of silica fume and superplasticiser (Bürge, 1982). w/c = 0.30.

8.4 Relationship between interface permeability and strength

Permeability and strength both depend on porosity and it is therefore natural for a correlation to exist between the two properties.

When casting the cement paste on sawn-cut surfaces of different aggregates, the water permeability of the interface is higher than that measured for the other constituents, paste and aggregate (ASTM, 1980). Experiments carried out on specimens such as those shown in Fig. 8.6 also revealed that the permeability at the aggregate–cement interface depends on the hydrostatic pressure, on the state of the aggregate's surface, on the length of curing of the specimens. As a matter of fact, it was ascertained that:

1. By gradually increasing the hydrostatic pressure on the interface, the water permeability increases accordingly;
2. Permeation is lower when the cement paste is cast on a broken surface rather than on a smooth surface of the same aggregate;
3. Prolonged water curing increases the paste–aggregate bond, strongly diminishes the permeability of the interface, while, in air-cured samples, a low hydrostatic pressure (approximately 2 atmospheres) is sufficient to detach the cement paste from the surface of the aggregate.

In all cases a higher permeability is associated with a lower paste–aggregate bond strength. This result, measured on the specimens referred to in Fig. 8.6, highlights a clear correlation between bond and permeability (Valenta, 1961).

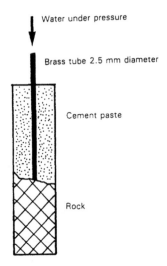

Fig. 8.6 Specimen for the permeability test of the aggregate–cement paste interfacial zone (Valenta, 1961).

8.5 Strength of concrete and interface bond strength

The strength of the paste–aggregate bond increases proportionally to the paste's compressive strength. This relationship results from the fact that both strengths diminish as the water/cement ratio increases. In other terms, the bond strength appears to depend on the same factors that determine the paste's strength.

Among these, an important role is played by porosity. For example if the same cement/aggregate ratio is used but the water/cement ratio is halved by means of the addition of superplasticizers, the compressive strength increases threefold, boosting from approximately 32 to 84 MPa (Valenta, 1961). In accordance with the reduction of the water/cement ratio, the total pore volume is halved as well as the pore volume of interface.

The densifying action of the admixture is also evidenced by the increase of pores having very small entrance radii (Sarkar and Aitcin, 1987).

The increase of strength as well as the reduction of porosity are accompanied by a heavy drop of the chloride diffusion. In this way, conductivity drops from 4600 coulomb to 150 coulomb passing from ordinary to high-strength concretes (Sarkar and Aitcin, 1987).

The increase of strength encountered in concrete prepared with a low water/cement ratio and the use of a superplasticizing admixture is probably due also to other reasons than a reduction in porosity.

In these concretes portlandite is in fact poorly crystallized and less abundant than in ordinary ones, and the aggregate is in direct contact with C-S-H instead of CH. Portlandite does not show a preferential orientation at the interface and the Ca/Si ratio of the hydrated calcium silicate is lower than the value usually found in the paste (Sarkar and Aitcin, 1987).

8.6 Morphology of the interfacial zone

All the properties of the interfacial zone which separates the aggregate from the bulk paste depend, directly or indirectly, on its morphology.

The interfacial zone, also called transition aura (Farran, Javelas, Maso and Perrin, 1972), has a composition and a microstructure that are quite different from the bulk paste and it is therefore hardly to be wondered if, although it is not greatly extended, it affects with its location the properties of the concrete. The morphology of this transition aura is determined - save for particular cases which will be described further below - by physical factors more than by chemical transformations. As a matter of fact, the transition aura shows a greater porosity (Farran, Javelas, Maso and Perrin, 1972), a more extended microcracking (Hsu, Slate, Sturman and Winter, 1963), and an increase of the comparatively larger portlandite crystals having a preferential orientation (Barnes, Diamond and Dolch, 1978; Farran, 1956; Grandet and Ollivier, 1980a; Moavenzadeh and Bremner, 1971). The characteristics of the interfacial zone also depend on the type of cement and it is therefore advisable to consider the different cases separately.

8.6.1 PORTLAND CEMENTS

It has by now been generally acknowledged that the cement paste has a different structure along the interface than the bulk paste. From a chemical point of view, the interfacial layer has a higher content of portlandite and ettringite and, from a physical viewpoint, it has a higher porosity. According to some Authors, a film of portlandite settles on the aggregate and is covered in its turn with a thin film of C-S-H (duplex film). This duplex film is 1 mm thick (Barnes, Diamond and Dolch, 1978; Barnes, Diamond and Dolch, 1979).

Beyond the duplex film, comparatively large and well formed crystals of Ca(OH) have been observed, with their C axis roughly parallel to the aggregate surface. The $Ca(OH)_2$ in contact with the aggregate is formed by crystals preferentially oriented with their C axis perpendicular to the surface of the aggregate. However, this duplex film has sometimes not been seen (Struble and Mindess, 1983; Scrivener and Pratt, 1986) and according to other investigations, in the initial stage of hydration a C-S-H film settles on the sand grains (Scrivener and Pratt, 1986). Conversely, it is generally admitted that the portlandite crystals are preferably oriented with their axes perpendicular to the interface (Grandet and Ollivier, 1980b), although well-defined hexagonal crystals are uncommon at the interface (Monteiro, Maso and Ollivier, 1985). The duplex film or C-S-H film are followed by a highly porous interface zone, approximately 50 mm deep, containing large portlandite crystals of different orientation, clusters of C-S-H and ettringite (Farran, Javelas, Maso and Perrin, 1972).

The thickness of the transition aura is practically unchanged in time as it depends only on the thickness of the liquid film which initially wetted the aggregate (Carles-Gibergues, Grandet and Ollivier, 1982).

With water/cement ratios of 0.5, also crystals of aluminates settle at random on the aggregate.

Within the transition aura the ettringite content diminishes regularly and considerably the farther one recedes from the interface and thus reaches an asymptotic value (Monteiro, Maso and Ollivier, 1985; Monteiro and Mehta, 1985) (see Fig. 8.7).

The differences in portlandite content between the transition aura and the paste are however very modest when one considers common concrete samples, submitted to normal mixing,

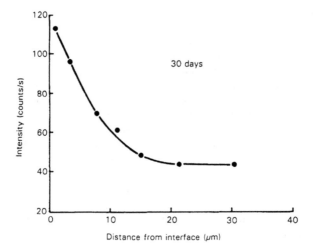

Fig. 8.7 Variation in ettringite content from aggregate surface (Monteiro, Maso and Ollivier, 1985).

instead of composite samples formed by casting the cement paste on large flat surfaces of aggregate (Scrivener and Gartner, 1988).

As regards the distribution of porosity, which is so closely linked to permeability, SEM examinations have shown that it diminishes regularly from the surface of the aggregate until it attains the average value existing in the bulk paste (see Fig. 8.8) (Scrivener and Pratt, 1987). The influence of the type of aggregate on the porosity distribution curve is perceptible but modest (Scrivener and Gartner, 1988). The lower density of the interfacial zone as

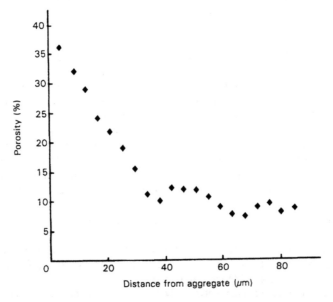

Fig. 8.8 Variation in porosity with distance from the aggregate surface (Scrivener and Pratt, 1987).

against that of the bulk paste has been confirmed by electric conductivity measurements. These measurements have also pointed up that the structure of the transition zone settles within 1–2 days (Xie Ping, Beaudoin and Brousseau, 1991). Calcite in calcareous aggregates may react with the C_3A of cement giving calcium carboaluminate hydrate $C_3A.CaCO_3.11H_2O$. The content of this compound drops rapidly and practically disappears at a distance of about 40 μm from the interface (Grandet and Ollivier, 1980c). The improved resistance to sulphates shown by concretes with calcareous aggregate as against those with siliceous aggregate was ascribed to the formation of this compound (Cussino and Pintor, 1972). This reaction, however, plays only a subordinate role in the paste–aggregate adhesion (Conjeaud, Lelong and Cariou, 1980).

A very strong bond is obtained when the concrete containing quartzite as aggregate is autoclaved (Tognon, Ursella and Coppetti, 1980). In these cases the transition aura becomes stronger than the bulk paste, because tobermorite crystals grow on the quartz surface epitaxically (Massazza and Pezzuoli, 1982).

8.6.2 BLENDED CEMENTS

Though few, data on the microstructure and morphology of the transition aura in concretes made of blended cements are in any case sufficient to show the influence of active additions such as fly ash, slag and silica fume.

The addition of fly ashes or slag to Portland cement reduces the degree of orientation of the portlandite crystals situated in the transition aura, but above all reduces the amount of $Ca(OH)_2$ formed in the proximity of the interface. The effect appears to be more marked for cements containing 60% slag than for cements containing 30% fly ashes (Saito and Kawamura, 1989).

The smaller quantity of crystals precipitated in the proximity of the interface can but have favourable effects on the ability of the interface zone to withstand aggressive agents (Saito and Kawamura, 1989).

The addition of 15% of microsilica produces a modest increase in both porosity and strength of the paste. In concrete, the addition slightly increases the porosity of the cement matrix but rises concrete strength by about 38% (see Table 8.4) (Scrivener, Bentur and Pratt, 1988).

This effect is due to a thickening of the paste in the transition aura, because (see Fig. 8.9 (Scrivener, Bentur and Pratt, 1988) the incorporation of the microsilica in the concrete dramatically reduces the porosity in this area. Fig. 8.9 shows that silica fume has practically eliminated the transition zone, because both the porosity and the portlandite are the same in the interfacial zone and in the cementitious matrix.

8.7 Durability

8.7.1 RESISTANCE TO CHEMICAL ATTACKS

The importance of the compactness of the interface zone with regard to the chemical aggression of concrete is obvious when one considers the relations existing between porosity, permeability and strength. Direct experimental data are few but sufficient to show that an improved adhesion between paste and aggregate is always associated with a higher resistance

Table 8.4 Composition and properties of concretes and cement pastes (Scrivener, Bentur and Pratt, 1988). Water/binder ratio 0.33.

	Concretes		Pastes	
Silica fume (%)	0	15	0	15
Cement content (kg/m³)	495	407	-	-
Total porosity at 28 days (volume %)	35	38	34	37
Compressive strength at 28 days (N/mm²)	77.9	107.6	84.5	85.6

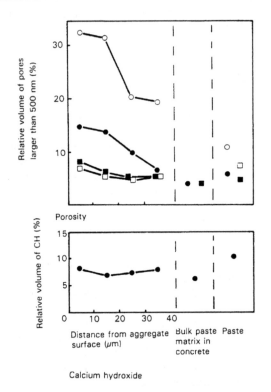

Fig. 8.9 Results of image analysis in the transition zone, in the bulk concrete matrix and in the pastes, for porosity and calcium hydroxide (Scrivener, Bwentur and Pratt, 1988).

to chemical attack.

The improved adherence of the paste to the aggregate, as a result of a chemical reaction, reduces the diffusivity and the penetration of aggressive ions into the concrete. Fig. 8.10 (Xie Ping and Ming-shu, 1988) shows that the expansion which takes place in pastes immersed in sodium sulphate solutions may be eliminated, or heavily reduced, by replacing quartz sand with sand formed with either Portland cement clinker or aluminous cement

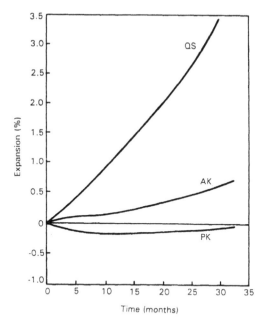

Fig. 8.10 Expansion of mortars stored in 5% (wt) Na$_2$SO$_4$ solution at 20°C (Xie Ping and Ming-Shu, 1988). QS = ground quartz sand; AK = ground aluminous clinker; PK = ground Portland clinker.

clinker, both reactive with respect to the cement paste (Xie Ping and Ming-shu, 1988).

The sulphate attack is less strong when a calcareous aggregate instead of a siliceous one is used. This favourable result was obtained with cements having both a low and a high Al$_2$O$_3$ content and has been ascribed to the reaction of C$_3$A with calcite and the formation of carboaluminate (Cussino and Pintor, 1972).

8.7.2 RESISTANCE TO FREEZING

The resistance of concrete to freeze-thaw cycles chiefly depends on the resistance to freezing of the interface zone, marked by a greater porosity and a higher content of large portlandite crystals as compared to the bulk paste.

In this case, too, direct data are few but in any case they are sufficient to draw up conclusions largely supported by the experimental practice. These conclusions may be summed up as follows (Valenta, 1961):

- the resistance of the cement paste–aggregate bond to freeze-thaw cycles is less than that of the paste and the aggregate;
- the resistance to cycles is greater when the aggregate has a broken surface rather than a smooth surface, that is to say, the greater the bond strength is;
- the resistance to freezing depends on the type of aggregate.

The results of the aforementioned tests show a rather high variability but are sufficient to indicate the important role played by the interface zone on the freeze-thaw resistance.

8.8 Conclusions

Direct experimental evidences of the influence of the environment on the interface existing between cement paste, aggregate and reinforcement are limited. Fortunately, the environmental factors affect the interfacial zone as the bulk concrete.

The most frequent damages occurring in concrete may be ascribed to volume variations induced by temperature changes, frost, growing of salt crystals, chemical reactions.

All these volume changes affect most the interface as it is generally the weakest part of the concrete.

The relative weakness of the interfacial layer is due to its higher chemical heterogeneity but above all to its porosity as compared to the bulk paste.

Higher porosity is accompanied by:

* lower strength,
* increased permeability to fluids,
* increased effective diffusion coefficient of species dissolved in pore solution,
* increased content of freezable water.

Microcracks, which are visible at the interface even in the absence of load, contribute to lowering the interface properties.

To protect interface from environmental attacks the same precautions must be taken as those which are necessary to obtain strong and durable concrete. That is to say, it is necessary to provide:

* a fairly high cement content taking into account the aggregate's grading,
* a tendentially low water/cement ratio,
* good compaction,
* thorough curing.

Besides these general precautions, durability can be improved by adopting special measures. For instance, the addition of pozzolanic material to the Portland cement reduces the permeability at the interface and seems to provide a simple means for reducing the environmental impact on concrete.

This favourable action appears to be connected with the lower amount of portlandite in the interface and its lower degree of orientation (Scrivener and Pratt, 1987) and lower porosity.

Another useful and viable provision appears to be the addition of superplasticizers to the fresh paste, that are capable of reducing the liquid film that covers initially the aggregate grains.

The bond could be effectively improved by autoclaving concretes containing quartz aggregates. Also in this case the strong increase in mechanical strength of interface is accompanied by a reduction in porosity and permeability but a generalization of the procedure finds considerable technical and financial obstacles.

8.9 References

Alexander, K.M., Wardlaw, J. and Gilbert, D.J. (1965) Aggregate–cement bond, cement paste strength and the strength of concrete, *The Structure of Concrete and its Behaviour under Load*, (eds

A.E. Brooks and K. Newman), Proceedings International Conference, London, September, Cement and Concrete Association, London, pp. 59–81.

ASTM (1980) Annual Book of ASTM Standards 1980, Part 14, C 234-71 and C 618-80.

Barnes, B.D., Diamond, S. and Dolch, W.L. (1978) The contact zone between Portland cement paste and glass "aggregate" surfaces, *Cement and Concrete Research*, Vol. 8, No. 2, pp. 233–44.

Barnes, B.D., Diamond, S. and Dolch, W.L. (1979) Micromorphology of the interfacial zone around aggregates in Portland cement mortar, *Journal of the American Ceramic Society*, Vol. 62, Nos 1-2, pp. 21–4.

Bartos, P.J.M., (ed.) (1982) *Bond in Concrete*, Proceedings of International Conference, Paisley, June, Applied Science Publishers, London.

Brooks, A.E. and Newman K. (eds) (1965) *The Structure of Concrete and its Behaviour under Load*, Proceedings of International Conference, London, September, Cement and Concrete Association, London.

Brunauer, S. (1966) The structure of hardened Portland cement paste and concrete, *8th Conference on the Silicate Industry*, Budapest, 9–12 June 1965, pp. 205–30.

Bürge, T.A. (1982) Densified cement matrix improves bond with reinforcing steel, *Bond in Concrete*, (ed. P.J.M. Bartos), Proceedings of International Conference, Paisley, 14–16 June 1982, Applied Science Publishers, London, pp. 32–5.

Carles-Gibergues, A., Grandet, J. and Ollivier, J.P. (1982) Evolution dans le temps de l'aureole de transition de pates contenant des ajouts, *International RILEM Colloquium, Liaisons Pâtes de Ciment/Matériaux Associés*, Toulouse, November, B–11-6.

Conjeaud, M., Lelong, B. and Cariou, B. (1980) Liaison pate de ciment Portland - granulats naturels, *Proceedings of the 7th International Congress on the Chemistry of Cement*, Editions Septima, Paris, Vol. III, VII-6 - VII-11.

Costa, U., Facoetti, M. and Massazza, F. (1990) Permeability of the cement–aggregate interface: influence of the type of cement, water/cement ratio and superplasticizer, *Admixtures for Concrete: Improvement of Properties*, E. Vazquez (ed.), RILEM International Symposium, Barcelona, May, Chapman & Hall, London, pp. 392–401.

Costa, U. and Massazza, F. (1988) Permeability and pore structure of cement pastes, 2nd International Conference on Engineering Materials, Bologna-Modena, 19–23 June.

Cussino, L. and Pintor, G. (1972) Indagine sul differente comportamento degli inerti silicico e calcareo nei conglomerati in funzione della composizione mineralogica del cemento, *Il Cemento*, Vol. 4, pp. 255–68.

Diamond, S. and Gomez-Toledo, C. (1978) Consistency, setting, and strength characteristics of a "low porosity" cement, *Cement and Concrete Research*, Vol. 8, No. 5, pp. 613–22.

Farran, J. (1956) Contribution mineralogique a l'etude de l'adherence entre les constituants hydrates des ciments et les materiaux enrobes, *Revue des Materiaux et Constructions*, No. 490–491, pp. 155–72; No. 492, pp. 191–209.

Farran, J., Javelas, R., Maso, J.C. and Perrin, B. (1972) Etude de l'aureole de transition existant entre les granulats d'un mortier et la masse de la pate de ciment hydrate, *Colloque International, Liaison de Contact dans les Materiaux Composites Utilises en Genie Civil*, Toulouse, November, Vol. I, pp. 60–76.

Grandet, J. and Ollivier, J.P. (1980a) Nouvelle methode d'etude des interfaces ciment-granulats, *Proceedings of 7th International Congress on Chemistry of Cement*, Paris, Editions Septima, Vol. III, VII-85 - VII-89.

Grandet, J. and Ollivier, J.P. (1980b) Orientation des hydrates au contact des granulats, *Proceedings of the 7th International Congress on the Chemistry of Cement*, Editions Septima, Paris, Vol. III, VII-63-8.

Grandet, J. and Ollivier, J.P. (1980c) Etude de la formation du monocarboaluminate de calcium hydrate au contact d'un granulat calcaire dans une pate de ciment Portland, *Cement and Concrete Research*, Vol. 10, No. 6, pp.759–70.

Hsu, T.T.C. and Slate, F.O. (1963) Tensile bond strength between aggregate and cement paste or mortar, *Journal of the American Concrete Institute, Proceedings,* Vol. 60, No. 4, pp. 465-86.

Hsu, T.T.C., Slate, F.O., Sturman, G.M. and Winter, G. (1963) Microcracking of plain concrete and the shape of the stress-strain curve, *Journal of the American Concrete Institute, Proceedings,* Vol. 60, No. 2, pp. 209-24.

Lyubimova, T.Yu. and Pinus, E.R. (1962) Crystallisation structures in the contact zone between aggregate and cement in concrete, *Kolloidnyi Zhurnal,* Vol. 24, No. 5, pp. 578-87.

Marsh, B.K., Day, R.L., Bonner, D.G. and Illston, J.M. (1983) The effect of solvent replacement upon the pore structure characterization of Portland cement paste, *Principles and Applications of Pore Structural Characterization,* J.M. Haynes and P. Rossi-Doria (eds), Proceedings RILEM/CNR International Symposium, Milan, 26-29 April, RILEM.

Massazza, F. and Pezzuoli, M. (1982) Reactions a l'interface entre pate de ciment et granulats dans des mortiers autoclaves, *International RILEM Colloquium, Liaisons Pâtes de Ciment/Matériaux Associés,* Toulouse, France, November, A-45-54.

Moavenzadeh, F. and Bremner, T.W. (1971) Fracture of Portland cement concrete structure, *Solid Mechanics and Engineering Design,* Southampton, April 1969, Vol. 2, pp. 997-100.

Monteiro, P.J.M., Maso, J.C. and Ollivier, J.P. (1985) The aggregate-mortar interface, *Cement and Concrete Research,* Vol. 15, No. 6, pp. 953-8.

Monteiro P.J.M. and Mehta, P.K. (1985) Ettringite formation on the aggregate-cement paste interface, *Cement and Concrete Research,* Vol. 15, No. 2, pp. 378-80.

Powers, T.C., Copeland, L.E., Hayes, J.C. and Mann, H.M. (1954) Permeability of Portland cement paste, *Journal of the American Concrete Institute, Proceedings,* Vol. 26, No. 3, pp. 285-98.

Powers, T.C., Copeland, L.E. and Mann, H.M. (1959) Capillary continuity or discontinuity in cement pastes, *Journal of PCA Research and Development Laboratories,* Vol. 1, No. 2, pp. 3-4.

Proceedings (1980) *7th International Congress on the Chemistry of Cement,* Paris, 30 June - 5 July 1980, Editions Septima, Paris, 1980.

Proceedings (1986) *8th International Congress on the Chemistry of Cement,* Rio de Janeiro, September 22-27, 1986, Secretaria Geral do 8°CIQC, Rio de Janeiro.

RILEM (1972) *Colloque International, Liaisons de Contact dans les Materiaux Composites Utilises en Genie Civil,* Toulouse, November.

RILEM (1982) *International Colloquium, Liaisons Pâtes de Ciment/Matériaux Associés,* Toulouse, 17-19 November, INSA, Toulouse, RILEM, Paris.

Roy, D.M. and Gouda, G.R. (1973) Porosity-strength relation in cementitious materials with very high strengths, *Journal of the American Ceramic Society,* Vol. 56, No. 10, pp. 549-50.

Saito, M. and Kawamura, M. (1989) Effect of fly ash and slag on interfacial zone between cement and aggregate, *Fly Ash, Silica Fume, Slag, and Natural Pozzolans in Concrete,* Trondheim, June, ACI SP-114, pp. 669-88.

Sarkar, S.L. and Aitcin, P.C. (1987) Comparative study of the microstructures of normal and very high-strength concretes, *Cement, Concrete, and Aggregates,* Vol. 9, No. 2, pp. 57-64.

Shah, S.P. and Slate, F.O. (1968) Internal microcracking, mortar-aggregate bond and the stress-strain curve of concrete, *The Structure of Concrete and its Behaviour under Load,* (eds A.E. Brooks and K. Newman), Proceedings of International Confereence, London, 28-30 September 1965, Cement and Concrete Association, London, pp. 82-92.

Scrivener, K.L., Bentur, A. and Pratt. P.L. (1988) Quantitative characterization of the transition zone in high strength concretes, *Advances in Cement Research,* Vol. 1, No. 4, pp. 230-7.

Scrivener, K.L. and Gartner, E.M. (1988) Microstructural gradients in cement paste around aggregate particles, *Bonding in Cementitious Composites,* (eds S. Mindess and S.P. Shah), Material Research Society, Vol. 114.

Scrivener, K.L. and Pratt, P.L. (1986) a preliminary study of the microstructure of the cement/sand bond in mortars, *Proceedings of 8th International Congress on Chemistry of Cement,* Rio de Janeiro, 22-27 September 1986, Vol. III, pp. 466-71.

Scrivener, K.L. and Pratt, P.L. (1987) The characterization and quantification of cement and concrete microstructures, *Porosite et Proprietes des Materiaux*, J.C. Maso (ed), Proceedings of RILEM International Conference, Versailles, September, Chapman & Hall, London, Vol. 1, pp. 61–8.

Struble, L. and Mindess, S. (1983) Morphology of the cement–aggregate bond, *Cement Composites and Lightweight Concrete*, Vol. 5, No. 2, pp. 79–86.

Tognon, G.P. and Cangiano, S. (1980) Interface phenomena and durability of concrete, *Proceedings of 7th International Congress on the Chemistry of Cement*, Paris, Vol. III, VII–133–8

Tognon, G.P., Ursella, P. and Coppetti, G. (1980) Bond strength in very high strength concrete, *Proceedings of 7th International Congress on Chemistry of Cement*, Paris, Editions Septima, Vol. III, VII-75-80.

Valenta, O. (1961) The significance of the aggregate–cement bond for the durability of concrete, *Durabilite des Betons*, RILEM Colloque International, Praha, Vol. I, pp. 53–87.

Wright, P.J.F. (1953) Entrained air in concrete, *Proceedings of Institution of Civil Engineers, Part I*, 2, No. 3, London, May, pp. 337–58.

Xie Ping, Beaudoin, J.J. and Brousseau, R. (1991) Flat aggregate–portland cement paste interfaces: I - electrical conductivity models, *Cement and Concrete Research*, Vol. 21, No. 4, pp. 515–22.

Xie Ping and Ming-shu, T. (1988) Effetto dell'interfaccia pasta di cemento Portland-aggregato sulla conduttivita' elettrica e sulla resistenza alla corrosione chimica della malta, *Il Cemento*, Vol. 1, pp. 33–42.

9

The effects of ageing on the interfacial zone in concrete

M.G. Alexander

9.1 Introduction

9.1.1 DEFINITION OF THE AGEING PROCESS

For the purposes of this chapter, the ageing process will be defined as:

"Continuing processes which alter the nature and microstructure of the interfacial zone".

These processes will usually:

- occur with time, but have rates which may be affected by temperature, pressure, etc.
- be related to aspects such as: the continuing hydration of cementitious or pozzolanic materials present in the zone; possible chemical reactions between cement paste and aggregate particles or other inclusions; deposition of crystalline products in the zone, for example due to ion migration; and possible leaching of products from the zone.

9.1.2 THE INTERFACIAL ZONE IN CONCRETE

Interfaces occur in composite cementitious materials in many forms: within the hydrating medium, between cement and pozzolanic admixtures, between cement and reinforcing steel or fibres, and between cement and aggregates (Mindess, 1989). This chapter will concentrate on the interfacial zone between cement paste and mineral aggregates such as mainly occur in concrete or mortar, and the influence of this zone and its ageing processes on the engineering properties of concrete.

Most structural grade concretes have mix proportions and production conditions that will result in the formation of a weak, porous interfacial zone between paste and aggregate. Even high strength concrete in which no mineral admixtures such as silica fume or fly ash have been used may exhibit such zones, as pointed out by Bentur and Cohen (1987). They showed that even at a water/cement ratio of 0.3, concretes based on Portland cement developed porous interfacial zones which were still evident after 28 days hydration.

For the sake of completeness, a brief review of the process of formation and nature of the interfacial zone in concrete is appropriate here. During mixing, casting, and consolidation (usually by vibration) of concrete, a layer of water accumulates around aggregate particles.

Interfacial Transition Zone in Concrete. Edited by J.C. Maso. RILEM Report 11.
Published in 1995 by E & FN Spon, 2-6 Boundary Row, London SE1 8HN. ISBN 0 419 20010 X.

Prior to initial set, this can be aggravated by additional bleed water gathering mainly under large aggregate particles. Furthermore, the so-called 'wall effect' (Bentur and Cohen, 1987; Mehta and Monteiro, 1988), prevents cement particles from packing efficiently around the aggregates. Thus, the aggregates are surrounded by a zone of relatively high water/cement ratio, in which conditions are markedly different from those in the bulk paste (Mehta and Monteiro, 1988). In summary, the nature of the zone in Portland cement concretes from the surface of aggregates outwards is (Scrivener and Pratt, 1986; Struble, 1988; Monteiro *et al.*, 1985):

1. A so-called 'duplex film' in direct contact with the aggregate surface, comprising CH preferentially oriented with c-axis normal to the aggregate surface, the outer part of the film being a layer of C-S-H gel (Barnes *et al.*, 1978). The duplex film is about 1 μm thick. (Note that not all investigators have detected this duplex film (Struble, 1988; Struble and Mindess, 1983).
2. A transition zone, or 'aureole de transition', up to 50 μm thick or more, which is rich in CH and ettringite, and also contains a large number of hollow shell hydration grains. The CH exists in large well-formed crystals with c-axes roughly parallel to the aggregate surface. A schematic sketch of the transition zone, given by Mehta (1986) is shown in Fig. 9.1.

Techniques for studying the interface zone were originally developed by Grandet and Ollivier (1980), Hadley (1972), Barnes (1975), and others.

The important physical properties of the interfacial region, relative to its effect on the bulk properties of concrete, are:

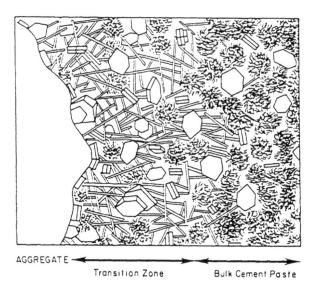

AGGREGATE ◄——————————►◄——————————►
Transition Zone   Bulk Cement Paste

Fig. 9.1 Diagrammatic representation of the transition zone and bulk cement paste in concrete (Mehta, 1986).

1. Increasing porosity gradients approaching the aggregate surface (Scrivener *et al.*, 1988).
2. Formation of preferential fracture planes due to the non-random orientation of CH which cleaves relatively easily along its basal plane (Scrivener and Pratt, 1986).
3. Inherent weaknesses of this zone which cause microcracks to form first here under any external stresses.

It has been suggested (Struble, 1988) that the large pore sizes in this zone may also represent critical Griffith-type flaws for initiation of fracture.

A further effect, not much mentioned in the literature, is the 'softening' effect of the transition zones when taken as a whole throughout a mass of concrete (Sarkar *et al.*, 1988). These zones are relatively compressible and also prevent proper interaction between paste and the (usually) stiff aggregates, thus lowering the stiffness of the concrete (Alexander, 1991).

Diamond *et al.* (1982), and Mindess and Diamond (1982), have shown that, in mortars, the minimum distance of separation between grains is, on average, less than 100μm. Thus, if the interfacial region is of the order of 50 μm wide, it appears that most of the concrete will be affected by this zone. However, it is likely that only the first 10 to 30 μm of this zone has substantially different physical properties from the bulk paste.

It may well be shown with further research that many of the well-documented changes and improvements in concrete properties with time (e.g. continuing strength gain, reducing rates of creep, improved impermeability) can be ascribed partially to changes in the interfacial zone, in addition to the more accepted changes in the paste phase. It might similarly be surmised that in silica fume concretes (say) where the interfacial zone appears to be very dense from the outset (Bentur and Cohen, 1987), age effects will be less.

The primary objective of this chapter will be to review data that show the effects of the interfacial zone on concrete during the ageing process defined above. This should help to focus attention on further research needs, and methods of manipulating the constituent materials and manufacturing processes of concrete to achieve a better construction material.

9.2 Effects during ageing

9.2.1 IMPLICATIONS OF THE INTERFACIAL ZONE FOR PROPERTIES OF CONCRETE

The general consensus of the literature seems to be that, for concrete, the bond between cement paste and aggregate exercises only a secondary influence on concrete strength (Mindess, 1988). Since the bond strength is governed by the properties of the interfacial zone, it follows that concrete strength may be relatively little affected by even large changes in the interfacial zone. This, however, may not be true for other properties of concrete such as stiffness and durability.

The presence of interfaces in concrete gives rise to at least three important questions, as Mindess (1988) has pointed out:

1. What effect do these interfaces have on the mechanical properties of concrete?
2. Is it possible to alter the interfaces in some controlled way to improve the properties of concrete?
3. Perhaps most importantly, what is the relative importance of the interfacial bonds in

determining strength, when compared to the effects of porosity?

To these three questions may be added a fourth:

4. Is it possible that the interfacial properties may influence properties of concrete other than strength to a more marked degree than strength?

The interfacial region is generally regarded as the 'weak link' in concrete, but attempts to improve the bond have generally had only a marginal influence on improving concrete compressive strength (Keru and Jianhua, 1988; Xueqan *et al.*, 1988; Jia *et al.*, 1986; Popovics, 1987). However, this is not to say that such improvements may not be of direct practical and economic importance - very often they are. For example, Alexander *et al.* (1968) showed that the flexural strength of concrete may be substantially improved by improving the bond strength, in their case by altering the type of aggregate.

Thus, an understanding of the interfacial zone and its properties is beneficial to improving the intelligent use of concrete. Nevertheless, we are still at a point where:

"While morphologies of various types of interfaces are reasonably well understood, their contributions to concrete properties are still largely a matter of conjecture." (Mindess, 1988).

It is proposed to deal with the effects during ageing as follows:

EFFECTS DURING AGEING

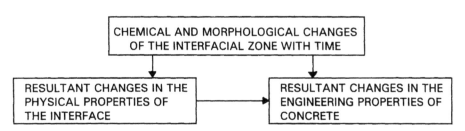

9.2.2 CHEMICAL AND MORPHOLOGICAL CHANGES OF THE INTERFACIAL ZONE WITH TIME

Starting with freshly mixed concrete in the plastic state, the interfacial zone changes from being predominantly a water-filled zone around an aggregate particle to a zone increasingly filled with solids and whose porosity reduces with time. Rates of deposition and growth of solids in this zone will differ from those in the bulk paste.

In the region of high local water/cement ratio around the aggregates, ions produced by the dissolution of cement grains accumulate by migration to form a solution supersaturated in calcium and hydroxyl ions. Thus the first effect is that of a precipitate of CH forming as a coating on the aggregate surface (Mehta and Monteiro, 1988). The next phase occurs when the sulphate, aluminate and silicate ions have also passed into the solution phase, and then, together with other calcium and hydroxyl ions, precipitate out as CH, ettringite, and C-S-H. The CH crystals tend to be large due to the large space available. Struble (1988), states that, in the interfacial or transitional zone, three general types of CH occur at various times:

1. A secondary layer of CH approximately 3 μm thick appears after a day or two of hydration. This layer is comprised of closely stacked CH crystals with (0001) faces oriented at an acute angle to the duplex film.
2. Larger, idiomorphic, tabular CH crystals develop at about the same time in the same region, many oriented with their c-axes parallel, rather than normal, to the aggregate surface.
3. Secondary CH accumulates throughout the transition zone during several weeks of hydration. This CH is often layered, tens of micrometers thick, variable in crystallographic orientation, and space-filling.

To a large degree, it is the presence of large quantities of CH that causes the weak mechanical strength of the interfacial region.

With the addition of mineral admixtures, in particular silica fume, the transition zone reduces markedly in thickness (Mehta and Monteiro, 1988; Carles-Gibergues *et al.*, 1982), and is far more homogeneous and dense. This has been graphically demonstrated by Bentur and Cohen (1987), whose photomicrographs are reproduced here in Figs 9.2 and 9.3. Fig. 9.2 shows a Portland cement mortar (w/c = 0.30) in which an initial (at 1 d) interfacial zone, characterized by voids and gaps, partially filled with CH crystals or porous hydration products (Fig. 9.2a) becomes far denser at 28 d (Fig. 9.2b). However, even at 28 d and w/c = 0.3, voids and gaps are still evident. In contrast, Fig. 9.3 shows a Portland cement/silica fume (sf) mortar (w/(c+sf) = 0.3; sf/c = 0.176) in which, even at 1 d (Fig. 9.3a) gaps in the interfacial zone are absent, and solid products form an intimate contact with the sand grain. There was no clear gradation of microstructure away from the grain surface. At 28 d, there is an extremely dense matrix extending up to and in contact with the sand grain (Fig. 9.3b).

 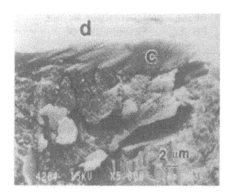

(a)   (b)

Fig. 9.2 Sand grain–paste interfacial zone of Portland cement mortars (Bentur and Cohen, 1987). The surface which was in contact with the sand grain is marked *d*: (a) 1 d old; (b) 28 d old, showing the CH rim (marked *c*).

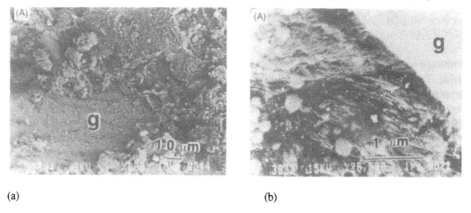

(a)   (b)

Fig. 9.3 Interfacial zone in Portland cement–silica fume mortars (Bentur and Cohen, 1987). The sand grain is marked *g*: (a) 1 d old; (b) 28 d old.

Carles-Gibergues *et al.* (1982) give results which show that the size of the transition zone does not change with time, although the orientation index of CH in the zone does change markedly with time. They also showed that fly ashes with open porosities or very fine materials like silica fume substantially reduce the size of this zone, while the use of fly ash from calcareous lignite or iron slag tends to increase the zone size.

Yuan and Odler (1987) concluded from their study of the interfacial zone between marble and tricalcium silicate paste that the higher CaO/SiO_2 ratio measured in the interfacial region was due to migration of Ca and OH ions into this zone. However, they indicated that a comparison of data obtained at different curing times revealed a gradual but only moderate increase of the CaO/SiO_2 ratio within the interfacial region with increasing time of hydration. Work of Scrivener and Pratt (1986) tends to confirm this slow increase in calcium content with time in the interfacial region. They state that the cleavage of CH tended to dominate fracture paths increasingly with time of hydration. They also note that in a very old concrete (85 y), CH which they believe nucleated during the early stage of hydration had persisted indefinitely.

Many authors have commented on chemical reactions occurring with time between the cement paste phase (including the pore solution) and aggregates of various types (Mehta and Monteiro, 1988; Yuan and Guo, 1987; Monteiro and Mehta, 1986b; Struble *et al.*, 1980; Maso, 1980). Many of these are claimed to be beneficial to the bond between paste and aggregate, but some authors point to chemical reactions actually weakening the bond strength. For instance, Yuan and Guo (1987) found that an excessive chemical dissolution of marble ($CaCO_3$) by the liquid phase of the adjoining cement paste may weaken the bond by precipitating large amounts of CH in the form of large crystals in the interface. Of course, the classic problems with alkali-aggregate reaction (AAR) stem from a relatively slow, on-going reaction between aggregates and highly alkaline cement pore solution. This can occur with both carbonate and siliceous aggregates and causes a decohesion of the cement-aggregate bond and cracking of the material (Hobbs, 1988; Regourd *et al.*, 1982). This subject is dealt with briefly later.

In general, there is a paucity of information in the literature on changes in the interfacial zone with time. Clearly, such data would depend on a great variety of factors, but might assist in helping to identify ageing processes.

9.2.3   RESULTANT CHANGES IN THE PHYSICAL PROPERTIES OF THE INTERFACE

While considerable work has been done on characterizing the morphological and, to a lesser extent, chemical and mineralogical nature of the interfacial zone, less data is forthcoming on physical properties of interfaces, and even less on variation of these properties with time. This is understandable, as it is generally more difficult to measure these properties on the relatively microscopic scale that is required, than to obtain optical and electron microscope images of this zone. However, the use of quantitative microscopy can give very useful data on physical properties such as porosity (Scrivener, 1988). A further complicating factor is that it is extremely difficult to measure physical properties such as bond strength or fracture toughness on a *real* interface, i.e. an interface in real concrete. Therefore, researchers often resort to testing *artificial* interfaces made by casting paste against rock surfaces prepared in different ways, and it is by no means clear how these results may differ from the real situation. There is no doubt that the two situations differ (Scrivener and Pratt, 1986; Scrivener and Gartner, 1988) and greater efforts must be made to test real interfaces.

9.2.3.1 Effect of ageing on bond strength
Bond strength is taken to refer to the cleavage or tensile-type adhesion of paste to an aggregate particle. Bond strength will depend on the common range of factors, i.e. type of cement and aggregate, surface preparation of the rock sample, age, temperature, type of cracking, and method of testing. Thus, quoted numerical values may not be directly comparable. Generally, bond strength increases with age, but not at the same rate as the cement paste itself.

Alexander *et al.* (1968) provided an admirable review and summary of bond strength data in 1965, and much of their work is still relevant today. They found that, in contrast to cement paste, bond strength (as a modulus of rupture) was dependent primarily on age, but not on the temperature of curing - see Fig. 9.4. Since paste strength is dependent on

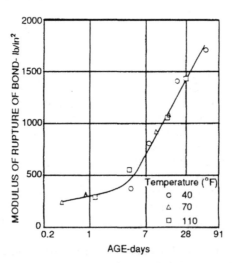

Fig. 9.4 Curve showing bond strength to be independent of curing temperature. w/c = 0.35 (Alexander *et al.*, 1968).

temperature, the ratio of bond to paste strength may vary widely with the temperature and duration of curing. At low temperatures (4–5°C), bond strength equalled or nearly equalled paste strength at all ages studied by Alexander and Taplin (1962 and 1964). At higher temperatures, paste strength exceeded bond strength at early ages, but at late ages bond strength and paste strength might again be equal - see Fig. 9.5. At this point (point A in Fig. 9.5), the mode of failure changes from preferential bond rupture to preferential failure in the paste. Giaccio and Zerbino (1986) also found that bond strength developed more slowly than paste strength, and that the bond strength was less sensitive than paste strength to variation in water/cement ratio of the paste. The results of Alexander *et al.* (1968) show the importance of long-term testing. Their work is also noteworthy for some findings that surface texture of prepared rock specimens had little influence on bond strength. They used broken, ground and sawn rock samples. However, they did state that naturally weathered rock samples tended to develop weak bond.

Massazza and Costa (1986), in a review article, quote data from Perry and Gillott (1983) on bond strength, shown in Fig. 9.6. For aggregates that are non-alkali-reactive, bond strength increases with time, the opposite occurring for alkali-reactive materials (glass, opal). The magnitude of the bond strength quoted is somewhat surprising. The same article also shows data from Zimbelmann (1985), given here in Fig. 9.7, which clearly shows the influence of aggregate type. In this case, bond strength values are considerably lower than tensile strength values for pastes or mortars.

Odler and co-workers (Yuan and Odler, 1987; Odler and Zürz, 1988) also studied the influence of aggregate type on cleavage (bond) strength using polished rock surfaces, and Portland cement or Portland cement/silica fume pastes. Their results are shown in Fig. 9.8. Clearly, silica fume enhances bond strength, especially of quartzite rock in this case. Failure in their cube samples occurred mainly between the 'duplex film' and the next region of ettringite and CH crystals. With silica fume, separation occurred preferentially within the paste, leaving significant amounts of hydrated material on the rock face. It was also shown that the bond of paste with marble was greater for opc paste than for pure C_3S paste (by a factor of about 3). The authors attributed this to the formation of carboaluminates in the

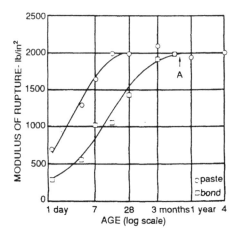

Fig. 9.5 The effect of age on bond and paste strength. w/c = 0.35 (Alexander *et al.*, 1968).

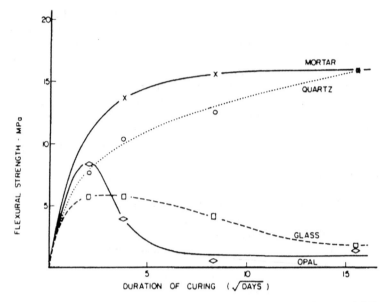

Fig. 9.6 Bond strength with time as affected by different aggregate types (Perry and Gillott, 1983).

reaction between $CaCO_3$ and C_3A, which they claim is beneficial for bond. Bond strength appeared to attain an ultimate value after about 1 month of curing.

Other workers also report 'improvements in bond strength' when using OPC/silica fume pastes, but give no quantitative details (Sarkar *et al.*, 1988; Larbi and Bijen, 1990).

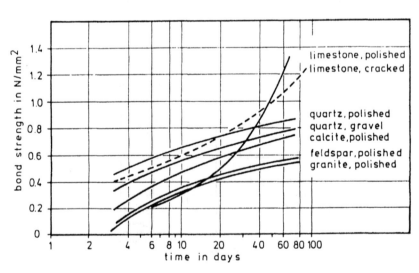

Fig. 9.7 Influence of aggregate type and surface preparation on bond strength (Zimbelmann, 1985).

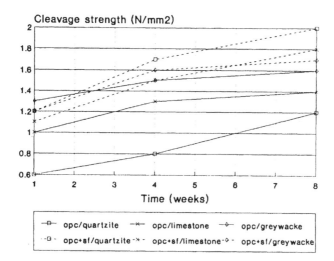

Fig. 9.8 Influence of cement and aggregate type on bond strength (Odler and Zürz, 1988).

9.2.3.2 *Effect of ageing on other physical properties*

(a) Fracture toughness
As noted by Struble (1988), it is difficult to estimate fracture toughness of specific microstructural regions, such as the interface, and few studies of the fracture process in the interfacial region have been carried out. Those references that were consulted had no data on the effect of ageing on fracture properties. Hillemeier and Hilsdorf (1977) measured the fracture toughness of cement–aggregate interfaces using wedge-loaded compact tension specimens, and found that the interface had significantly lower toughness than either paste or aggregate. Typical values of K_{Ic} were 0.3–0.5 MPa$\sqrt{}$m for paste, and 0.1–0.2 MPa$\sqrt{}$m for interfaces, higher if the interfaces had been 'modified' with a polymer admixture. Keru and Jianhua (1988), using single edge-notched beam tests, reported fracture energy values for interfaces of 20–35 J/m^2, depending on pretreatment of the aggregate. Alexander (1991) reported K_{Ic} values of 0.5–0.8 MPa$\sqrt{}$m , and fracture energy values of 7–20 J/m^2 for interfaces, using chevron-notched cylindrical beam specimens. These fracture values were very similar to values obtained for pure paste.

A number of authors (Bentur and Cohen, 1987; Mindess and Diamond, 1982; Zhang *et al.*, 1988) have pointed out that fracture in the interfacial zone often occurs not immediately at the interface, but at a distance from it (typically 1–30 μm depending on the nature of the interface).

(b) Porosity
Scrivener, Bentur and Pratt (1988), using quantitative microscopy, measured porosity changes of ordinary and silica fume concrete in the interfacial zone with time. Their results are shown in Fig. 9.9. The densification of the interfacial zone with time due to continuing deposition of hydration products is clearly apparent, as is also the densifying effect of silica fume, even from a very early age.

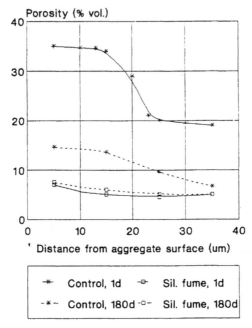

Fig. 9.9 Effect of cement type, and age on porosity of the interfacial zone (based on data in Scrivener, Bentur and Pratt, 1988).

Kayyali (1987) tested hardened cement paste, small crushed aggregates, and a micro-concrete made from these constituents for porosity distribution using MIP. His object was to endeavour to determine indirectly the porosity of the interfacial zone. As expected, total porosity reduced with age. What is of interest in the results, however, is the lower porosity of the micro-concrete in comparison with the paste. Using calculations based on mix proportions, Kayyali also showed that the composite's porosity was significantly lower than would be predicted from the known values for paste and aggregate. He concluded that the porosity of the interface must therefore be lower than that of the paste matrix, so much so that the aggregate was 'masked' by the interface and was not intruded. His first conclusion is very questionable, based on numerous studies showing higher porosity of the interface, but it is possible that a 'duplex film' or something similar may prevent aggregates from contributing to the overall porosity of a composite.

(c) Microhardness
Lyubimova and Pinus (1962) originally provided data on microhardness in 1962, showing the interface to be softer than either aggregate or bulk paste, and inferring from the data an interfacial width of 30–40 μm. More recently, Mehta and Monteiro (1988) used a Vickers Pyramid to measure microhardness (H) at the interfacial zone, and showed that between 30 d and 1 year of age, H increased from 6.3 to 7.9 kg/m^2. Of importance was also their observation that the 1 year interfaces exhibited radial corner cracks on testing which did not occur for the 30 d specimens, indicating a harder but more brittle interface with age. Mehta and Monteiro also suggest the use of this technique to measure directly the fracture toughness of the interface, using a relationship given by Anstis *et al.* (1981).

9.2.4 RESULTANT CHANGES IN THE ENGINEERING PROPERTIES OF CONCRETE

There is little data in the literature linking interfacial ageing effects to changes in concrete properties with time. Therefore what follows is an attempt to draw together some threads in order to promote a greater awareness of this important topic, and to stimulate more work on it. Ageing effects are not always directly referred to.

9.2.4.1 Strength
Whether there is a direct link between interfacial bond strength and strength of concrete has been the subject of much dispute. The following review by Struble *et al.* (1980) is appropriate here.

"Some investigations have shown a relationship between bond strength and concrete strength. For example, Alexander *et al.* (1968) reported a linear correlation (by multiple regression analysis) between concrete strength (compressive and tensile) and paste and bond strengths, where the coefficient of paste strength was approximately double that of the bond strength". (See later). "Conversely, Fagerlund (1973) concluded that the cement–aggregate bond is not important to concrete compressive strength, by theory or by experiment.

"There is evidence of considerable microcracking at the cement–aggregate interface during loading of concrete. Microcracking and its growth rate have been incorporated into several conflicting hypotheses explaining observed effect or lack of effect of cement–aggregate bond on concrete strength. Hsu *et al.* (1963) found that the ultimate concrete strength was not strongly dependent on bond strength and suggested that increasing the bond strength would increase the stress level at which extensive microcracking began. Similar conclusions were reported by Perry and Gillott (1983), who found little effect of bond strength on compressive strength above the stress level at which bond cracks are initiated. Scholer (1967), on the other hand, suggested that the stress level for initiation of microcracks is a function primarily of mortar strength. Based on measurements of microcracking in concrete during compressive tests, he hypothesized that the cement–aggregate bond influenced concrete strength by controlling the amount of microcracking necessary to reach failure. Finally, Patten (1972) suggested that, at high stress levels, a poor bond allows cracks to propagate more rapidly, hastening failure. His suggestion explained observed decreases in bond strength when a silicone release agent was used to eliminate chemical adhesion between mortar and coarse aggregate."

The consensus from most studies would seem to indicate that increasing the bond strength does increase concrete strength, by the order of 20% to 40% going from 'no bond' to 'perfect bond'. Alexander *et al.* (1968) gave a simple relationship between bond strength and either compressive or flexural strength, based on regression analysis - see Fig. 10 (based on Mindess, 1989). In Fig. 9.10, the 'no bond' situation has $\sigma_f(\text{bond}) = 0$, the 'perfect bond' situation has $\sigma_f(\text{bond}) = \sigma_f(\text{paste})$. Clearly, a change in compressive or flexural strength of the paste has about twice as much effect on concrete strength than an equivalent change in bond strength. In contrast to these results, work by Nepper-Christensen and Nielsen (1969), in which bond between mortar and an artificial aggregate of glass marbles was reduced by coating the marbles with a thin layer (12 μm) of soft plastic, showed that the effect of the coating was to reduce the compressive strength of the 'concrete' by up to 2.5 times.

Bentur and Cohen (1987) were able to show that by improving the density of the interfacial zone, and hence presumably increasing the bond, by using silica fume, improvements in

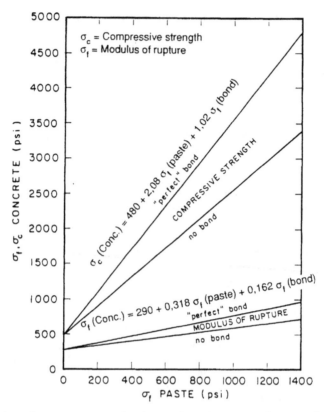

Fig. 9.10 Effect of cement–aggregate bond strength on the compressive strength (σ_c) and flexural strength (σ_f) of concrete (taken from Mindess, 1989, based on Alexander and Taplin, 1962 and 1964).

concrete strength of the order of 25% to 30% over plain cement concrete were realized. This occurred without any corresponding increase in the strength of the respective pastes, pointing to the improvements in the interfacial region as therefore being the most likely source of improved concrete strength. The same results and conclusions were arrived at by Sarkar *et al.* (1988), who found concrete compressive strength increases for silica fume concrete over plain concrete of about 20% for similar water/cement ratios (0.21 to 0.24). As Aitcin (1988) has pointed out, by densifying the interface and thus improving the interaction between paste and aggregate, concrete for the first time begins to act as a truly composite material.

Scrivener and Gartner (1988), from their own work and that of others, suggest that most of the strength enhancement in concrete is due to improved paste–aggregate bonding. This would naturally lead to the premise that, due to ageing effects, there comes a point where changes in paste strength contribute very little to improved concrete properties, while concurrent changes in the interfacial zone may contribute significantly to these improvements. It can also be suggested that it is not so much improved bond *strength* that is important with ageing, but improved *densification* of the interfacial zone which will help to reduce stress-raising flaws and voids. Bentur (1988) also concurs with this view, stating that attempts to improve bond 'per se' may not necessarily enhance concrete properties much, since it is the entire transition zone that influences mechanical behaviour. This would be one explanation why Popovics (1987) found very little strength increase in concretes in which he improved the direct bond with aggregates, but did not necessarily alter the characteristics of the interfacial zone as a whole. Clearly these hypotheses require considerable extra work to be done on them.

Aggregates in their own right exercise an influence on concrete strength, as shown conclusively many years ago by Kaplan (1959). What has not been quantified, however, is the relative contributions to enhanced strength of factors such as aggregate type, strength and elastic modulus, shape, surface texture (both macro and micro), and influence of the interfacial zone. Kaplan concluded that the two most important aggregate factors for concrete compressive strength were elastic modulus and surface texture. Massazza and Costa (1986) appear to agree by stating that the adhesion between paste and aggregate essentially depends on the aggregate roughness. They quote results from Cottin *et al.* (1982) where the surface roughness of a calcareous aggregate was made to vary from smooth to acid-etched to crushed grains, with improvements in compressive strength.

Alexander (1991) also found an appreciable influence of aggregate type on concrete compressive strength, using a broad range of South African aggregates. Typical results are shown in Fig. 9.11. It appears from these results that aggregate elastic modulus does not necessarily play an important role - the very low stiffness siltstone aggregate concrete had high strength.

9.2.4.2 Stiffness

Elastic modulus of concrete can be represented by simplified models, two of which are shown schematically in Fig. 9.12 – the Series Model (for a composite soft material, such as typically when normal weight aggregates are used), and the Parallel Model (for a composite

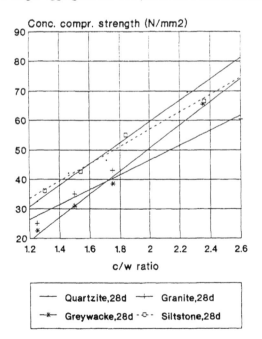

Fig. 9.11 Influence of aggregate type on concrete compressive strength at 28 d (Alexander, 1991).

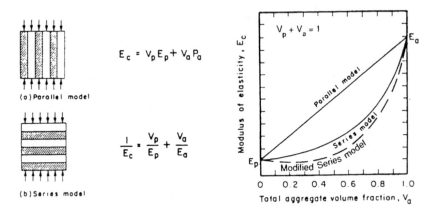

Fig. 9.12 Possible models for elastic modulus of concrete.

hard material, such as typically when lightweight aggregates are used). The effect of soft, porous interfaces would be virtually negligible for the Parallel Model, but may be very appreciable for the Series Model. Depending on the average thickness and relative stiffness of the interfacial zones, it could be postulated that a relationship such as that shown as the 'Modified Series Model' could result, where the interfaces markedly reduce the overall stiffness of the composite. Obviously, if ageing has the effect of densifying the initially porous interface zones, then the material stiffness can be considerably enhanced with age.

Such a model might also be used to explain why different aggregate types, even though of similar stiffness, can produce concretes of markedly different elastic modulus, as found by Alexander (1991). A selection of results is shown in Figs 9.13 and 9.14. Fig. 9.13 shows that, while aggregate elastic modulus may exercise a general influence on concrete elastic modulus, there are many anomalies in the results (such as higher modulus concrete being achieved with lower modulus aggregate), which may more easily be explained in terms of interface effects. Fig. 9.14 refers directly to ageing effects, showing that from 28 days to 6 months, concretes may achieve considerable gain in stiffness, to an extent greater than would be predicted on the basis of the 28 day relationships and strength gains over the given period of time. This is shown schematically in Fig. 9.15, where the relative contributions to concrete stiffness between 28 days and 6 months of additional matrix (paste) strength, and surmised interface effects are indicated. The relative influence of the interface reduces with age and increasing strength for the dolomite concrete, while that for the andesite concrete is virtually constant. Ageing effects here can be deduced as densifying the interface with time, thereby reducing the extent and increasing the stiffness of these zones. This argument was also used by Aitcin (1988) in explaining the remarkably high elastic modulus (> 50 GPa) of very high strength concrete (> 120 MPa) made with a water content of only 120 l/m^3.

In contrast possibly to the results given above, Sarkar *et al.* (1988) give relationships for stiffness of concretes based on ordinary Portland cement, and Portland cement + silica fume, at very low water/cement ratios (0.21 to 0.27). Results covering 7 d, 28 d and 1 year are shown in Fig. 9.16. The silica fume concretes have achieved significantly higher strength and stiffness than the opc concretes. However, the elastic modulus – strength relationships for

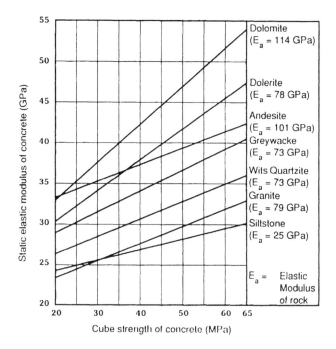

Fig. 9.13 Influence of aggregate type on concrete elastic modulus (Alexander, 1991).

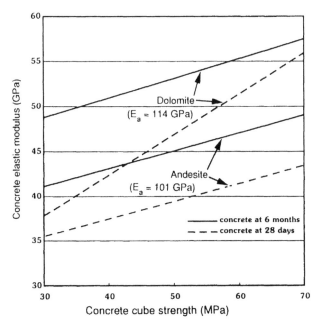

Fig. 9.14 Increase in concrete elastic modulus with age for two different aggregate types (Alexander, 1991).

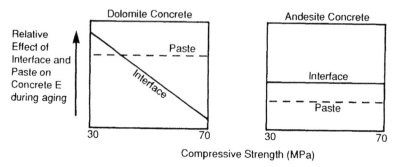

Fig. 9.15 Schematic representation of relative effects of interface and paste on concrete stiffness during ageing.

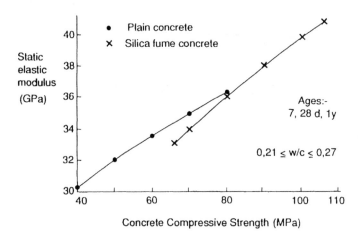

Fig. 9.16 Influence of cement type on elastic modulus of concrete (based on data in Sarkar *et al.*, 1988).

the two concretes can be taken as falling in a common band, in which case the silica fume concretes are not inherently stiffer for any given strength. The authors mention that a large (1–2 mm) porous zone existed around aggregate particles for the OPC concretes, but the interfacial zone for the silica fume concretes was dense right up to the aggregate surface. Unfortunately, matrix and aggregate elastic moduli were not given, thus making it difficult to assess the influence of the interfacial zones. However, it would appear that, for these very low water/cement ratio concretes, ageing effects may be less.

9.2.4.3 Brittleness
Mindess (1989) has noted that one major consequence of increasing the cement–aggregate bond strength (or, indeed, the concrete strength in general) is that the brittleness of the concrete also increases. This leads to a decrease in the fracture energy, probably because the stress level at which extensive microcracking begins in the interfacial zone is increased. In the general case, this is not very significant, since it is the reinforcing steel which imparts

any real 'ductility' to reinforced concrete structures. Ageing in this case may be though of as a 'detrimental' process, but much work remains to be done in this area.

9.2.4.4 Durability

Early work by Valenta (1969) indicated a much higher permeability of the interface than that of aggregate or paste, but later work by Wakeley and Roy (1982) and by Mindess *et al.* (1985), showed little or no significant influence of the interface on concrete permeability. This is probably due to the relatively small volume of the interfacial region in concrete.

However, Xueqan *et al.* (1987) in a study of concrete manufactured by first mixing the aggregates with a low water/cement ratio slurry, followed by vacuum de-watering after placing, showed that durability could be improved by also preactivating the aggregates to allow them to form stable hydrates in the interface, rather than CH. Some of their results are shown in Fig. 9.17 with regard to expansion in a sulphate solution. They attribute the improved concrete properties directly to an improved interfacial region.

If the processes used by Xueqan *et al.* can be regarded as an accelerated ageing process, then the question arises as to whether improvements in durability of normal concretes with time may be due primarily to improvements due to ageing in the interfacial zone.

Alkali-aggregate reaction as a detrimental ageing process

Alkali-aggregate reaction (AAR) can be regarded as a form of interfacial interaction between the products of cement hydration and mineral aggregates. It does not fall within the scope of this chapter to provide a comprehensive treatment of AAR. However, it will be briefly covered since it represents an important ageing process in regard to interfacial phenomena. AAR can comprise one of three forms: alkali-silica reaction (ASR), alkali-silicate reaction, and alkali-carbonate reaction (see Hobbs, 1988).

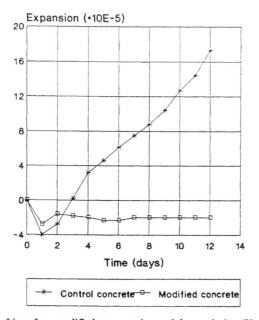

Fig. 9.17 Expansion of interface-modified concrete in a sulphate solution (Xueqan *et al.*, 1987).

For the purposes of this chapter no distinction will be drawn between these three, since they all produce a deterioration in concrete properties. The primary effects of AAR are: (a) interfacial cracking between cement paste and aggregate; (b) mortar matrix cracking; (c) often severe cracking of concrete structures; (d) overall expansion of concrete members; (e) occasionally structural inadequacy. An example of an interfacial crack is shown in Fig. 9.18.

While the effects of AAR may be visually alarming, the actual performance of AAR-affected structures is often predictable if based on laboratory tests of affected concrete, and structural safety is often not seriously compromised (Blight and Alexander, 1985). The simple reason for this is the enormous degree of stress redundancy represented by concrete in compression. However, the interfacial effects of AAR have very marked effects on the mechanical properties of concrete. This is invariably a time-dependent effect, due to the nature of AAR. A summary is given here.

(a) Strength. Reduction in strength of concrete depends on the level of expansion. Hobbs (1988) gives data that show compressive and tensile strength reductions of up to 25% and 30%, respectively, for concretes with unrestrained expansion below 0.5%. Blight and Alexander (1985) state that the strength of concrete disrupted by AAR is decreased to a surprisingly small degree, of the order of 20%. Hobbs also notes that concretes, once cracked, actually continue to gain in strength with time after the initial strength reduction.

(b) Stiffness. Hobbs (1988), and Blight and Alexander (1985) show data indicating elastic modulus reduction in AAR-affected concrete of between 20% and 60%. This is primarily due to the cracked nature of the concrete, and the presence of compressible gel between cement matrix and aggregate. Creep strains measured on AAR affected concrete were about 2 to 4 times as large as creep strains in undeteriorated concrete which was otherwise similar (Blight and Alexander, 1985).

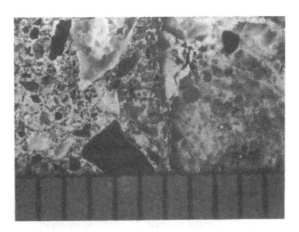

Fig. 9.18 Interfacial crack in concrete due to AAR.

Tests on a large overhead motorway in South Africa showed that there was only a very nominal decrease (approximately 5%) in overall stiffness of the structure over a 6 year period (Blight *et al.*, 1989). This was despite crack widths on the surface increasing from an average of about 1 mm in 1982 to 3–4 mm in 1988.

9.3 Modification of the interfacial zone to alter the properties of concrete

Ageing has been defined in this chapter as being those continuing processes which alter the nature and microstructure of the interfacial zone. The idea is that these processes will usually occur due to containing hydration, ion migration and solids deposition in the interfacial zone with time. These continuing processes will affect the concrete properties with time, often beneficially. As Bentur (1988) has pointed out, in special cement composites where very high strength or toughness is needed, the interfacial characteristics and their influence on the properties of the composite very often play a crucial role. Therefore, we must know how to control and modify these characteristics to achieve optimum results.

There are several processes or design approaches that can be used to alter the nature of the interfacial zone in order to improve concrete properties. These modification processes can cause a more rapid alteration of the zone, or a fundamental change in the nature of the zone, and can therefore be thought of as either an acceleration in ageing or radical change in the zone so that ageing effects are minimal. It is not within the scope of this chapter to deal comprehensively with these modification processes, but for completeness, they are briefly covered below.

Modification processes chiefly have to do with *densification* of the interfacial zone, *grain refinement*, and improving the *physico-chemical* interaction between cement paste and aggregates. They may often overlap in practical situations.

9.3.1 DENSIFICATION PROCESSES

Densification is typically achieved by using a very fine mineral admixture such as silica fume in the mix. Such fine powders appear to have at least four effects: (a) they improve the packing density in the interfacial region, thereby largely eliminating the 'wall effect' mentioned earlier; (b) they act to reduce bleeding, thus also reducing the size of the transition zone (Carles-Gibergues *et al.*, 1982); (c) they act as growth nuclei for multiple generation of CH crystals which therefore have smaller size (i.e. grain refinement); and (d) they participate in longer-term pozzolanic reactions which continue to densify the interfacial zone with time. Numerous studies have experimentally pointed to the above conclusions (Bentur and Cohen, 1987; Sarkar *et al.*, 1988; Jia *et al.*, 1986; Odler and Zürz, 1988; Larbi and Bijen, 1990; Monteiro and Mehta, 1986a). Obviously, densification processes often occur simultaneously with grain refinement processes.

9.3.2 GRAIN REFINEMENT PROCESSES

Grain refinement can be achieved in different ways. Monteiro and Mehta (1986a, 1986b) report on two processes (other than the use of silica fume mentioned above): one using an ASTM Type K (expansive) cement, the other using carbonate aggregates. The Type K cement favours grain refinement in the interfacial zone by replacing large CH crystals (having a preferential orientation) with small CH crystals having random orientation, and apparently leads to improved concrete strength. The use of carbonate aggregates favours grain refinement by a process of dissolution of large CH crystals and of calcium carbonate to produce a network of smaller crystals of a basic calcium carbonate.

9.3.3 PHYSICO-CHEMICAL INTERACTION PROCESSES

These processes often involve pre-treating or pre-activating aggregates with chemical solutions, or low water/cement ratio pastes, thus supposedly giving them a greater affinity for reacting or interacting with the cement paste. The intention is primarily to improve the bond in the interfacial zone. Work has been reported by Xueqan *et al.* (1988), 1987)), Keru and Jianhua (1988), Popovics (1987), Yuan and Guo (1988), and others. Reported improvements in concrete strength are of the order of 10% to 30%. Popovics (1987) on the other hand, testing concretes made with aggregates pre-coated with thin epoxy layers impregnated with either fine sand or unhydrated Portland cement particles found no real improvement in strength. However, in his case it is probable that no significant improvement in the interfacial zone as a whole was effected. Another process is that of mixing aggregates with a low water/cement ratio paste as the initial step in concrete mixing, before adding the bulk of the water. This helps in reducing bleeding and improving interfacial bond (Xueqan *et al.*, 1988; Xueqan *et al.*, 1987; Hayakawa and Itoh, 1982).

Another well-known process of altering the bond in the interfacial zone is autoclaving. Massazza and Costa (1986), reviewing work by Tognon *et al.* (1980) and Massazza and Pezzuoli (1982), report that autoclaving only slightly increases strength of concrete made with calcareous and basaltic aggregate, but enhances it considerably when granite, jasper or quartz aggregates are used. They ascribe this to chemical reactions occurring between silica and cement paste, enhancing the cement–aggregate bond. Microhardness measurements across the interface bear this out. Autoclaving was also found to increase the elastic modulus of concretes with siliceous aggregates, but reduce it when calcareous aggregates were used.

9.4 Concluding remarks

There is a growing appreciation for the role that the interfacial zone between paste and aggregate plays in the engineering behaviour of concrete. Points that need to be stressed are as follows:

1. Changes in microstructure and density of the interfacial zone with time have an important influence on the properties of concrete, for example increased strength, stiffness, durability, and brittleness.
2. Bond strength between paste and aggregate may be less important than originally surmised. Evidence that failure occurs at some finite distance from the interface suggests that it is the density and cohesive strength of the interfacial zone itself that may be of greater importance.
3. There is a research need to link microstructural studies of the interfacial zone with determinations of the resultant engineering properties of concrete. There is also a need to study changes in the interfacial zone with time, with reference to different types of

cements and aggregates, temperature etc. There is a lack of useful data on this at present.

4. Other opportunities for useful practical research exist in studying the possibilities of 'engineering' the interfacial zone in order to modify concrete properties beneficially. The processes involved here will involve densification, grain refinement, and other physico-chemical effects.

9.5 References

Aitcin, P.C. (1988) From gigapascals to nanometres, *Advances in Cement Manufacture and Use*, (ed. E.M. Gartner), Engineering Foundation, New York, pp. 105–113.

Alexander, K.M. and Taplin, J.H. (1962) Concrete strength, paste strength, cement hydration and the maturity rule, *Australian Journal of Applied Science*, Vol. 13, No. 4, pp. 277–84.

Alexander, K.M. and Taplin, J.H. (1964) Analysis of the strength and fracture of concrete based on an unusual insensitivity of cement-aggregate bond to curing temperature, *Australian Journal of Applied Science*, Vol. 15, No. 3, pp. 160–70.

Alexander, K.M., Wardlaw, J. and Gilbert, D.J. (1965) Aggregate–cement bond, cement paste strength and the strength of concrete. *The Structure of Concrete and its Behaviour under Load*, (eds A.E. Brooks and K. Newman), Proceedings of International Conference, London, Cement and Concrete Association, pp. 59–81.

Alexander, M.G. (1991) Fracture energies of interfaces between cement paste and rock, and application to the engineering behaviour of concrete, *Fracture Processes in Brittle Disordered Materials*, (eds J.G.M. van Mier *et al.*), Proceedings International RILEM Conference, Noordwijk, Netherlands, E & FN Spon, London, pp. 333–46.

Anstis, G.R., Chantikul, P, Lawn, B.R. and Marshall, D.B. (1981) A critical evaluation of indentation techniques for measuring fracture toughness. Part 1: Direct crack measurements. Part 2: Strength method. *Journal of the American Ceramic Society*, Vol. 64, No. 9, pp. 533–38, and 539–43.

Barnes, B.D. (1975) Morphology of the Paste–Aggregate Interface. Ph.D Thesis, Purdue University.

Barnes, B.D., Diamond, S. and Dolch, W.L. (1978) The contact zone between Portland cement paste and glass "aggregate" surfaces, *Cement and Concrete Research*, Vol. 8, No. 2, pp. 233–44.

Bentur, A. (1988) The role of the interface in controlling the performance of high quality cement composites, *Advances in Cement Manufacture and Use*, (ed. E.M. Gartner), Engineering Foundation, New York, pp. 227–37.

Bentur, A. and Cohen, M.D. (1987) Effect of condensed silica fume on the microstructure of the interfacial zone in Portland cement mortars, *Journal of the American Ceramic Society*, Vol. 70, No. 10, pp. 738–43.

Blight, G.E. and Alexander, M.G. (1985) Damage by alkali-aggregate reaction to reinforced concrete structures made with Witwatersrand quartzite aggregate and examples of repair methods, *Concrete Beton*, Concrete Society of South Africa, Vol. 10, pp. 14–23.

Blight, G.E., Alexander, M.G., Ralph, T.K. and Lewis, B.A. (1989) Effect of alkali-aggregate reaction on the performance of a reinforced concrete structure over a six-year period, *Magazine of Concrete Research*, Vol. 41, No. 147, pp. 69–77.

Carles-Gibergues, A., Grandet, J. and Ollivier, J.P. (1982) Evolution of the "Aureole de transition" with ageing in blended cement pastes, *International RILEM Colloquium, Liaisons Pâtes de Ciment/Matériaux Associés*, Toulouse, France, B.11-B.16 (in French).

Cottin, B., Marcdargent, S. and Cariou, B. (1982) Reactions between active aggregates and hydrated cement paste, *International RILEM Colloquium, Liaisons Pâtes de Ciment/Matériaux Associés*, Toulouse, France, C.20-C.26, (in French).

Diamond, S., Mindess, S. and Lovell, J. (1982) On the spacing between aggregate grains in concrete and the dimension of the Aureole de Transition, *International RILEM Colloquium, Liaisons Pâtes de Ciment/Matériaux Associés*, Toulouse, France, C.42-C.46.

Fagerlund, G. (1973) Strength and porosity of concrete, *Proceedings of International Symposium on Pore*

Structure and Properties of Materials, RILEM/IUPAC, Prague, D51-D73.

Giaccio, G. and Zerbino, R. (1986) Factors affecting cement paste–aggregate bond, *Proceedings of the 8th International Congress on the Chemistry of Cement,* Rio de Janeiro, Vol VI, pp. 331-3.

Grandet, J. and Ollivier, J.P. (1980) New method for the study of cement–aggregate interfaces, *Proceedings of the 7th International Congress on the Chemistry of Cement,* Editions Septima, Paris, Vol III, VII - 85-89 (in French).

Hadley, D.H. (1972) The Nature of the Paste–Aggregate Interface, Ph.D Thesis, Purdue University.

Hayakawa, M. and Itoh, Y. (1982) A new concrete mixing method for improving bond mechanisms, *Bond in Concrete,* (ed. P.J.M. Bartos), Applied Science Publishers, London, pp. 282-8.

Hillemeier, B. and Hilsdorf, H.K. (1977) Fracture mechanics studies on concrete compounds, *Cement and Concrete Research,* Vol. 7, No. 5, pp. 523-35.

Hobbs, D.W. (1988) *Alkali-Silica Reaction in Concrete,* Thomas Telford, London.

Hsu, T.T.C., Slate, F.O., Sturman, G.M. and Winter, G. (1963) Microcracking of plain concrete and the shape of the stress-strain curve, *Journal of the American Concrete Institute, Proceedings,* Vol. 60, No. 2, pp. 209-224.

Jia, W., Baoyuan, L. Songshan, X. and Zhongwei, W. (1986) Improvement of paste–aggregate interface by adding silica fume, *Proceedings of the 8th International Congress on the Chemistry of Cement,* Rio de Janeiro, Vol. III, pp. 460-5.

Kaplan, M.F. (1959) Flexural and compressive strength of concrete as affected by the properties of coarse aggregates, *Journal of the American Concrete Institute, Proceedings,* Vol. 55, pp. 1193-207.

Kayyali, O.A. (1987) Porosity of concrete in relation to the nature of the paste–aggregate interface, *Materials and Structures,* Vol. 20, No. 115, pp. 19-26.

Keru, W. and Jianhua, Z. (1988) The influence of the matrix–aggregate bond on the strength and brittleness of concrete, *Bonding in Cementitious Composites,* (eds S. Mindess and S.P. Shah), Materials Research Society, Vol. 114, pp. 29-34.

Larbi, J.A. and Bijen, J.M. (1990) Orientation of calcium hydroxide at the Portland cement paste–aggregate interface in mortars in the presence of silica fume: a contribution, *Cement and Concrete Research,* Vol. 20, No. 3, pp. 461-70.

Lyubimova, T.Yu. and Pinus, E.R. (1962) Crystallisation structure in the contact zone between aggregate and cement in concrete, *Kolloidnyi Zhurnal,* Vol. 24, No. 5, pp. 578-87. In Russian.

Maso, J.C. (1980) The bond between aggregates and hydrated cement paste, *Proceedings of the 7th International Congress on the Chemistry of Cement,* Editions Septima, Paris, Vol. I, VII - 1/3-15.

Massazza, F. and Costa, U. (1986) Bond: paste–aggregate, paste–reinforcement and paste–fibres, *Proceedings of the 8th International Congress on the Chemistry of Cement,* Rio de Janeiro, Vol I, pp. 158-80.

Massazza, F. and Pezzuoli, M. (1982) Interface reactions between cement paste and aggregates in autoclaved mortars, *International RILEM Colloquium, Liaisons Pâtes de Ciment/Matériaux Associés,* Toulouse, A.45-A.54.

Mehta, P.K. (1986) *Concrete: Structure, Properties and Materials.* Prentice-Hall, Englewood Cliffs, N.J.

Mehta, P.K. and Monteiro, P.J.M. (1988) Effect of aggregate, cement, and mineral admixtures on the microstructure of the transition zone, *Bonding in Cementitious Composites,* (eds S. Mindess and S.P. Shah), Materials Research Society, Vol. 114, pp. 65-76.

Mindess, S. (1988) Bonding in cementitious composites: how important is it? *Bonding in Cementitious Composites,* (eds S. Mindess and S.P. Shah), Materials Research Society, Vol. 114, pp. 3-10.

Mindess, S. (1989) Interfaces in concrete, *Materials Science of Concrete,* (ed. J. Skalny), The American Ceramic Society, pp. 163-80.

Mindess, S. and Diamond, S. (1982) a device for direct observation of cracking of cement paste or mortar under compressive loading within a SEM, *Cement and Concrete Research,* Vol. 12, No. 5, pp. 569-76.

Mindess, S., Gray, R.J. and Skalny, J.P. (1985) Effects of silica fume additions on the permeability of hydrated Portland cement paste and the cement–aggregate interface, *Proceedings, ACI-RILEM Symposium on Technology of Concrete when Pozzolans, Slags and Chemical Admixtures are Used,* Monterrey, Mexico, pp. 121-42.

Monteiro, P.J.M., Maso, J.C. and Ollivier, J.P. (1985) The aggregate–mortar interface, *Cement and*

Monteiro, P.J.M., Maso, J.C. and Ollivier, J.P. (1985) The aggregate–mortar interface, *Cement and Concrete Research*, Vol. 15, No. 6, pp. 953–8.

Monteiro, P.J.M. and Mehta, P.K. (1986a) Improvement of the aggregate–cement paste transition zone by grain refinement of hydration products, *Proceedings of the 8th International Congress on the Chemistry of Cement*, Rio de Janeiro, Vol. III, pp. 433–7.

Monteiro, P.J.M. and Mehta, P.K. (1986b) Interaction between carbonate rock and cement paste, *Cement and Concrete Research*, Vol. 16, No. 2, pp. 127–34.

Nepper-Christensen, P. and Nielsen, T.P.H. (1969) Modal determination of the effect of bond between coarse aggregate and mortar on the compressive strength of concrete, *Journal of the American Concrete Institute, Proceedings*, Vol. 66, pp. 69–72.

Odler, I. and Zürz, A. (1988) Structure and bond strength of cement–aggregate interfaces, *Bonding in Cementitious Composites*, (eds S. Mindess and S.P. Shah), Materials Research Society, Vol. 114, pp. 21–8.

Patten, B.J.F. (1972) The effects of adhesive bond between course aggregate and mortar on the physical properties of concrete. UNICIV Report No. R-82, Univ. of New South Wales, Australia.

Perry, C. and Gillott, J.E. (1983) The long term decrease in cement–aggregate bond strength of a quartzite aggregate containing trace amounts of iron, *Durability of Building Materials*, Vol. 1, No. 4, pp. 305–11.

Popovics, S. (1987) Attempts to improve the bond between cement paste and aggregate, *Materials and Structures*, Vol. 20, No. 115, pp. 32–8.

Regourd, M., Hornain, H., Mortureux, B., Poitevin, P. and Peuportier, H. (1982) Alteration of the cement paste-aggregate bond in concrete - the alkali-aggregate reaction, *International RILEM Colloquium, Liaisons Pâtes de Ciment/Matériaux Associés*, Toulouse, France, B.17–B.25, in French.

Sarkar, S.L., Diatta, Y. and Aitcin, P.C. (1988) Microstructural study of aggregate/hydrated paste interface in very high strength river gravel concretes, *Bonding in Cementitious Composites*, (eds S. Mindess and S.P. Shah), Materials Research Society, Vol. 114, pp. 111–6.

Scholer, C.F. (1967) The role of mortar-aggregate bond in the strength of concrete, *Highway Research Record*, 210, pp. 108–17.

Scrivener, K.L. (1988) Quantification of microstructure, *Advances in Cement Manufacture and Use*, (ed. E.M. Gartner), Engineering Foundation, New York, pp. 3–13.

Scrivener, K.L., Bentur, A. and Pratt, P.L. (1988) Quantitative characterisation of the transition zone in high strength concretes, *Advances in Cement Research*, Vol. 1, No. 4, pp. 230–7.

Scrivener, K.L., Crumbie, A.K. and Pratt, P.L. (1988) A study of the interfacial region between cement paste and aggregate in concrete, *Bonding in Cementitious Composites*, (eds S. Mindess and S.P. Shah), Materials Research Society, Vol. 114, pp. 87–8.

Scrivener, K.L. and Gartner, E.M. (1988) Microstructural gradients in cement paste around aggregate particles, *Bonding in Cementitious Composites*, (eds S. Mindess and S.P. Shah), Materials Research Society, Vol. 114, pp. 77–86.

Scrivener, K.L. and Pratt, P.L. (1986) A preliminary study of the microstructure of the cement/sand bond in mortars, *Proceedings of the 8th International Congress on the Chemistry of Cement*, Rio de Janeiro, Vol. III, pp. 466–71.

Struble, L. (1988) Microstructure and fracture at the cement paste-aggregate interface, *Bonding in Cementitious Composites*, (eds S. Mindess and S.P. Shah), Materials Research Society, Vol. 114, pp. 11–20.

Struble, L. and Mindess, S. (1983) Morphology of the cement–aggregate bond, *International Journal of Cement Composites and Lightweight Concrete*, Vol. 5, No. 2, pp. 79–86.

Struble, L., Skalny, J. and Mindess, S. (1980) A review of the cement–aggregate bond, *Cement and Concrete Research*, Vol. 10, No. 3, pp. 277–86.

Tognon, G.P., Ursella, P. and Coppetti, G. (1980) Bond strength in very high strength concrete, *Proceedings of the 7th International Congress on the Chemistry of Cement*, Editions Septima, Paris, Vol. II, VII-75 to VII-80.

Valenta, O. (1969) Durability of concrete, *Proceedings of the 5th International Congress on the Chemistry of Cement*, Cement Assoc. of Japan, Tokyo, Vol. 3, pp. 193–225.

Wakeley, L.D. and Roy, D.M. (1982) A method for testing the permeability between grout and rock, *Cement and Concrete Research*, Vol. 12, No. 4, pp. 533–4.

Xueqan, W., Dongxu, L., Qinghan, B., Liqun, G. and Minshu, T. (1987) Preliminary study of a composite process in concrete manufacture, *Cement and Concrete Research*, Vol. 17, No. 5, pp. 709–14.

Xueqan, W., Dongxu, L., Xiun, W. and Minshu, T. (1988) Modification of the interfacial zone between aggregate and cement paste, *Bonding in Cementitious Composites*, (eds S. Mindess and S.P. Shah), Materials Research Society, Vol. 115, pp. 35–40.

Yuan, C.Z. and Guo, W.J. (1987) Bond between marble and cement paste, *Cement and Concrete Research*, Vol. 17, No. 4, pp. 544–52.

Yuan, C.Z. and Guo, W.J. (1988) Effects of bond strength between aggregate and cement paste on the mechanical behaviour of concrete, *Bonding in Cementitious Composites*, (eds S. Mindess and S.P. Shah), Materials Research Society, Vol. 114, pp. 41–7.

Yuan, C.Z. and Odler, I. (1987) The interfacial zone between marble and tricalcium silicate paste, *Cement and Concrete Research*, Vol. 17, No. 5, pp. 784–92.

Zhang, X., Groves, G.W. and Rodger, G.A. (1988) The microstructure of cement aggregate interfaces, *Bonding in Cementitious Composites*, (eds S. Mindess and S. P. Shah), Materials Research Society, Vol. 114, pp. 89–96.

Zimbelmann, R. (1985) A contribution to the problem of cement–aggregate bond, *Cement and Concrete Research*, Vol. 15, No. 5, pp. 801–8.

INDEX

s UK
Group UK Ltd.
71024
9B/585